TEXTBOOK OF FOREST MENSURATION

THE AUTHORS

Prof. (Dr.) C.S. Bohra has more than 19 years of teaching, research and administrative experience at various capacities in various universities and research organizations in the field of Agriculture sciences, Forestry, Agroforestry, Forest Ecology & Biodiversity conservation, Soil Science, Plantation Forestry, Medicinal Herbs, Climate Change, Carbon sequestration, Environmental Science and Disaster management. He started his journey in 2003-04 from Kumaun University, Nainital (Uttarakhand) just after qualifying National Eligibility Test (NET) from UGC/ICAR and worked with many other organizations of International repute in the last more than nineteen years of his career like, Govind Ballabh Pant University of Agriculture & Technology, Pantnagar, Uttarakhand, Kumaun University, Nainital, Graphic Era Hill University, Dev Bhoomi Uttarakhand University, Dehradun, Quantum University, Roorkee, Uttarakhand & other affiliated colleges from H. N. B. Garhwal Central University, Srinagar, Uttarakhand & Sri Dev Suman Uttarakhand University, Uttarakhand. Earlier, he was associated with many funding agencies like University Grants Commission (UGC), New Delhi, National Oilseeds & Vegetable Oil Development Board (NOVOD), Gurgaon, Govt. of India, National Mission on Bamboo Application (NMBA), Govt. of India, New Delhi and Department of Science & Technology (DST), Govt. of India. Dr. Bohra is the one & only PDF recipient of UGC's Dr D S Kothari scheme in all over the country in agroforestry domain. Professor Bohra has published five books, many chapters and more than two dozen of research papers and articles in the International/National journals of repute. He is also linked with many journals as member of editorial board and reviewer. Recently, Dr. Bohra has also been conferred by various agencies as 'BEST RESEARCH AWARD', 'YOUNG SCIENTIST AWARD-2022' and 'HARIT KRANTI AWARD-2022'. In the past years, many conferences have been organized by Dr. Bohra and also presented his research in various National/International conferences/symposiums. Currently, Dr. Bohra is associated as Professor & Dean in the College of Agriculture & Environmental Technology, with Surajmajl University, Kichha (U.S. Nagar), Uttarakhand, India.

Dr. Deepti Bisht Bohra is an accomplished Assistant Professor with a strong background in Geography/Environmental Geography with about 11 years of academic experience in Higher Education. She has made significant contributions into Geographical research especially on Dairy farming and Animal Husbandry aspects of the region. She holds a Doctorate degree in the field of Geography from Kumaun University, Nainital, where she conducted ground-breaking research on different economic and environmental aspects and prospective of dairy farming and animal husbandry in the region. As a seasoned educator, Dr. Deepti has taught and mentored countless students at undergraduate level. She is known for active, collaborative and project-based teaching-learning approach, which has led to deeper understanding with creativity on students. Dr. Deepti's research interests span a wide range of topics in Geography, including Environmental Geography, Economic Geography, Indian Geography, Rural Geography, Urban Geography and Tourism Geography. She is a recipient of research fellowship from Indian Council of Social Science Research (ICSSR, New Delhi) for her research work in the field of Animal Husbandry. She has been conferred by 'Young Scientist Award-2022' for her contribution in the field of geography. She has published extensively, authoring peer-reviewed research papers/articles in collaboration with other leading experts in the field. Her work has been recognized and cited by peers, contributing to the advancement of knowledge in the field of rural and urban geography. Dr. Deepti is dedicated to fostering a passion for Environmental & Rural Geography among her students and peers. She continues to explore new avenues of research and innovation, striving to make meaningful contributions to environmental, rural as well as urban geographical aspects and inspire the next generation of scholars. Presently, Dr. Deepti Bisht Bohra is associated as an Assistant Professor with the Department of Higher Education, Government of Uttarakhand, India at Lal Bahadur Shashtri Government Post Graduate College, Halduchaur (Nainital), Uttarakhand.

TEXTBOOK OF FOREST MENSURATION

(For B.Sc. Agriculture/ B.Sc. Forestry students/ FRO/Forest Guards/Forester/ ACF Trainees/ Competitive Examinations)

Dr. C. S. Bohra

Professor & Dean

College of Agriculture & Environmental Technology
Surajmal University, Kichha (U.S. Nagar)
Uttarakhand - 263148, India

&

Dr. Deepti Bisht Bohra

Assistant Professor

Lal Bahadur Shashtri Government Post Graduate College Halduchaur,
Nainital, Uttarakhand-263142, India

2025

Daya Publishing House®
A Division of
Astral International Pvt. Ltd.
New Delhi – 110 002

ISBN: 978-93-5919-500-1

Published by : **Daya Publishing House®**
A Division of
Astral International Pvt. Ltd.
– ISO 9001:2015 Certified Company –
4736/23, Ansari Road, Darya Ganj
New Delhi-110 002
Ph. 011-43549197, 8130496929
e-mail: info@astralint.com
Website: www.astralint.com

Printed at : **Replika Press Pvt Ltd**

Dedicated to

Our Bundle of Joy, Our Lifeline;

Baby Navya

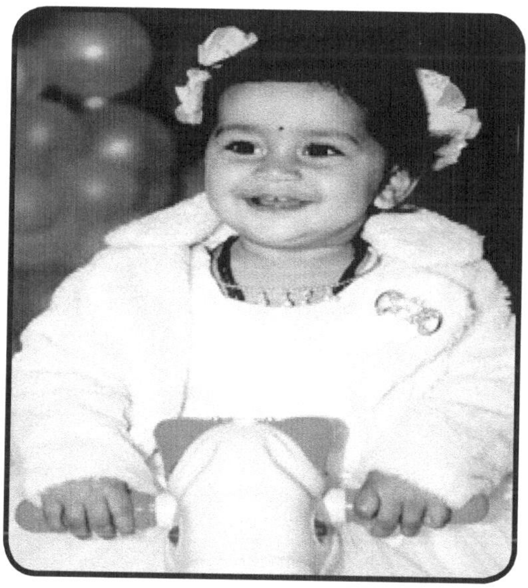

Preface

Professionals from multifarious organizations has created significant competition among the graduates of various subjects. The success of any professional lies with the assured employment in government, NGOs and industrial sectors. Forestry subject is of no exception. Being a narrow science, but offered almost several SAUs & private universities distributed across the country, it has created large number of Agriculture and forestry graduates, which resulted in competition and compelled the agriculture and forestry graduates to acquire up to date information in their subject for getting successful induction in public and private sectors. For being successful, one must be thorough in each and every aspect of the subject which demand a comprehensive book to resolve the issues.

Various competitive examinations conducted by state and central government organizations particularly UKPSC (Forest Ranger, ACFs, PCS), HPSC, UPSC (Preliminary exams), ARS (JRF, SRF and Scientist) and Bank exams, etc. along with the B.Sc. Agriculture (H.) and B.Sc. Forestry (H.) are predominantly conducted through objective and subjective systems. For objective examination wide range of books are available for the use by all competitions. However, for subjective type examinations, only very few books are available which again have not compiled the readymade and concrete information for each and every aspect of the subject. The huge subjects often pose the threat to the examinations which resulted in consuming more time and money in search of the actually relevant materials.

To mitigate this problem, the present book on Forest Mensuration has been conceived and designed to benefit the Forestry & Agriculture graduates who are preparing for the examinations like, UKPSC (Forest Ranger, Forest guards, PCS), HPSC, UPSC (Preliminary exams), ARS (JRF, SRF and Scientist) and Bank exams., etc along with the regular honors graduate degrees in agriculture and forestry. This hand bank on Forest Mensuration provides the knowledge on the subject and extends adequate hands-on experience through practice with well-defined diagrams and Fig. s.

It encompasses several facets of forest mensuration subject extensively which included Introduction, Measurement of Trees, Measurement of Tree Forms, Forest Sampling and Inventory, Objective questions and answers along with other additional and related information on forest mensuration in the form of glossary.

Present book covers wide range of syllabus and problems with appropriate explanations which will be a ready reckoned and perfect book for wide range of competitive examinations. Adequate care has been taken to avoid errors. In Agriculture and Forestry sector, this is a maiden attempt and bound to happened with errors. Such errors if any, noticed by readers of this book can be informed to the author for necessary corrections and publications in all their future editions. Extending useful and constructive criticism and suggestions are most welcome for augmenting the contents in future.

Prof. (Dr.) C. S. Bohra

Dr. Deepti Bisht Bohra

Contents

UNIT 1

Introduction

Forest Mensuration, deals with the quantification of forests, trees, and forest products. We can distinguish in it techniques for direct or indirect measurement, estimation procedures using statistical relationships, and methods of prediction where the variable time takes part.

	Measurement (Direct, Indirect)	*Estimation (Statistical)*	*Prediction (Over Time)*
Logs (Products)	Length, diameter Cubication (Volume) Log rules Defect, Quality	Volume functions Sawn timber conversion Stacked wood	
Trees	DBH, height, bark Cubication Stem analysis	Volume functions Taper functions Bark functions Product assortment	
Stands	Stand tables	Height-DBH DBH distributions Volume functions Inventories	Site quality Growth Mortality

Definition and Scope of Forest Mensuration

Forest

- An area set aside for the production of timber and other forest product.
- A plant community predominantly of trees and other woody vegetation usually with a closed canopy (Glossary).
- Forests are the lands of more than 0.5 ha with a tree canopy cover of more than 10% which are not primarily under agricultural or urban land use (FAO, 2000).

Mensuration

- It is derived from Latin word mensura which means measure. It means measurement of length, mass and time etc.
- It is an art and science of locating, measuring and calculating the length of lines, areas of planes, and volumes of solids
- It is that branch of mathematics which is concerned with the determination of lengths, areas and volumes.

Forest Mensuration

- Forest Mensuration deals with the determination of the volume of logs, trees, and stands, and with the study of increment and yield (Graves, 1906).
- Forest Mensuration is that branch of forestry which deals with the determination of dimensions (eg. Diameter, height, volume etc), form, age and increment of single trees, stands or whole woods, either standing or after felling (Chaturvedi and Khanna, 1986)
- Forest Mensuration is the determination of dimensions, form, weight, growth, and age of trees individually or collectively, and of the dimensions of their products (Helms, 1998).
- It is a tool that provides facts about the forest crops or individual trees to sellers, buyers, planners, managers and researchers.

Objectives

Forest mensuration provides quantitative information regarding forest resources that will allow making reasonable decisions on its density, use and management.

Forest mensuration serves the following objects:

- Basis for sale
- Basis for management
- Measurement for research
- Measurement for planning

Scope

- It is the branch of forestry which provides foundations of measurement principles applicable to any forest management problems (Fig. 1)
- Has a wide scope.

- ☆ Involves all stakeholders i.e. Labors, buyers, sellers, contractors, planners, managers/foresters and researchers.
- ☆ Applicable to any forest measurement problems of wildlife management, watershed management, insect and disease incidence, recreation, tourism and in fact, many of the mensurational aspects of multiple use forestry.
- ☆ Forest Mensuration is the application of measurement principles to obtain quantifiable information for forest management decision making.
- ☆ The application of statistical theory and use of electronic computer for data processing have brought about revolutionary changes in forest measurement problems.
- ☆ Forest mensuration should make full use of these tools but its principles must be based on sound biological knowledge.

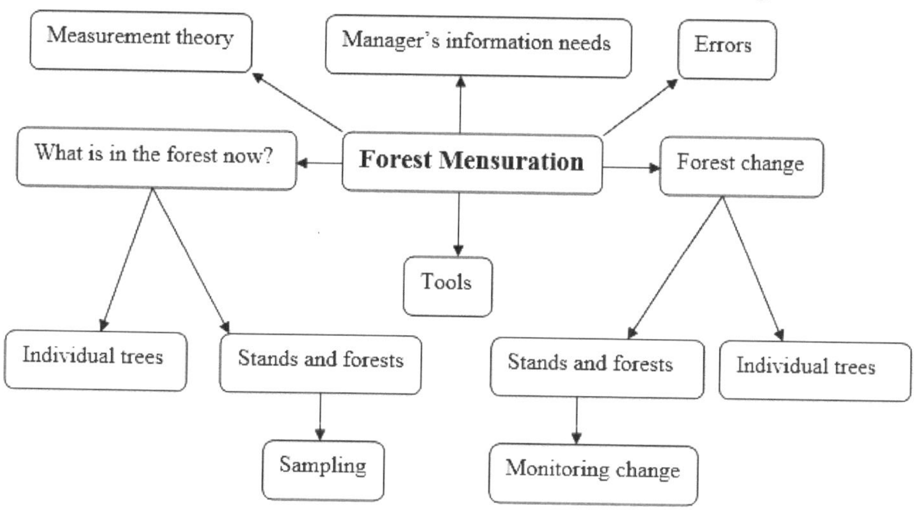

Fig. 1: Scope of Forest Mensuration

Importance of Forest Mensuration

It is the keystone in the foundation of forestry. Forestry in the broadest sense is a management activity involving forestland, the plants and animals on the land, and the human as they use the land. Forest mensuration solves the following questions which justifies its importance in forestry.

- ☆ What silvicultural treatment will result in best regeneration and growth?

- What species is most suitable for reforestation?
- Is there sufficient timber to supply a forest industry and for an economic harvesting operation?
- What is the value of timber and land?
- What is the recreational potential?
- What is the wildlife potential?
- What is the status of biodiversity on the area?
- What is the status of the forest as a carbon sink?
- What is the status of forest now?
- How is the forest changing?
- What can we do to manage the forest properly?
- How can it be assessed?
- And for what purpose?

It helps to answer all these questions and concepts involved in forest management. "You can't efficiently make, manage, or study anything if you don't locate and measure".

If you cannot measure it, you cannot manage it and good management decisions require good data; thus, mensuration has vital role in forest management. Forest mensuration is concerned with the obtaining of information about forest resources. The ultimate objective of forest mensuration is to provide quantitative information regarding this resource that will allow making reasonable decisions on its density, use and management.

Bias, Accuracy and Precision

- The difference between a measurement and the true value of the quantity measured is the true error of the measurement, and is never known since the true value of the quantity is never known.
- A discrepancy is the difference between two measured values of the same quantity, it is not an error.

Sources of Errors

- Instrumental
- Personal
- Natural- temperature, humidity, gravity, wind, magnetic declination

Kinds of Errors

- ☆ Mistake
- ☆ Systematic errors (cumulative errors)
- ☆ Accidental errors (compensating errors)

Bias

- ☆ It refers to the systematic errors that may result from faulty measurements procedures, instrumental errors, flaws in the sampling procedure, errors in the computations, mistakes in recording and so on.

Common Sources of Bias

- ☆ Flaw in measurement instrument or tool, e.g. survey tape 50 cm short
- ☆ Flaw in the method of selecting a sample, e.g. some observers always count the boundary trees while others always exclude it
- ☆ Flaw in the technique of estimating a parameter, e.g. stand volume: using a volume function or model in a forest without prior check of its suitability for application in that forest; inappropriate assumptions about the formulae and
- ☆ Subjectivity of operators etc.

How to Minimize Bias

- ☆ Continual check of instruments and assumptions
- ☆ Meticulous training
- ☆ Care in the use of instruments and application of methods

Accuracy

- ☆ Accuracy is the degree of perfection obtained. It is the closeness of a measurement to the true value.
- ☆ It is the success of estimating the true value of a quantity
- ☆ It refers to the size of the deviation of a simple estimate from the true population
- ☆ The ultimate objective is to obtain accurate measurement.

Accuracy Depends on

- ☆ Precise instruments
- ☆ Precise methods
- ☆ Good planning

Precision

- ☆ Precision is the degree of perfection used in the instruments, the methods and the observations. It is the degree of agreement in a series of measurements.
- ☆ It is the closeness of a measurement to the average value
- ☆ It refers to the deviation of sample values about their mean.
- ☆ It is also used to describe the resolving power of a measuring instrument or the smallest unit in observing a measurement. In this sense, the more decimal places used in a measurement, the more precise the measurement.

In sampling, accuracy refers to the size of the deviation of a sample estimate from the true population value. Precision, expressed as a standard deviation, refers to the deviation of sample values about their mean, which, if biased, does not correspond to the true population value. It is possible to have a very precise estimate in that the deviations from the sample mean are very small; yet at the same time, the estimate may not be accurate if it differs from the true value due to bias.

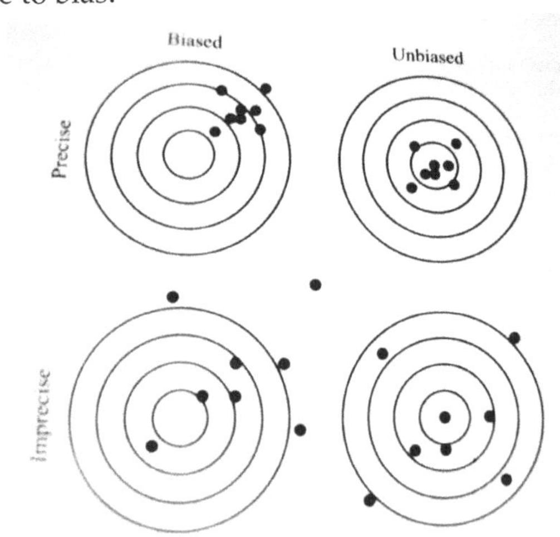

Fig. 2: Precision, bias and accuracy of a target shooter. The target's bull's eye is analogous to the unknown true population parameter, and the hole represent parameter estimates based on different samples. The goal is accuracy, which is the precise, unbiased target (Adapted from Shiver and Borders, 1996).

For example: One might carefully measure a tree diameter repeatedly to the nearest millimeter, with a caliper that reads about 5mm low. The results of this series of measurements are precise because there is little variation between readings, but they are biased and inaccurate because of faulty adjustment of the instrument. The relation between accuracy A, bias B, and precision P can be expressed as $A_2 = B_2 + P_2$. This means that if we reduce B_2 to zero, accuracy equals precision

Accuracy in Forest Mensuration

- Though mensuration is the branch of mathematics, forest mensuration does not attempt to secure absolute accuracy.
- Forest mensuration aims at reasonable or relative accuracy, i.e. maximum accuracy which is profitable and possible to obtain in practice. For the following reasons, foresters are compelled to be content with relative accuracy.
- Characteristics of trees
- Varying methods and conditions of felling and conversion
- Instruments and conditions in which they are used
- Personal bias of the estimator
- Biological character of the forest
- The use to which the measurements are to be put
- Cost

UNIT 2

Measurement of Trees

Diameter Measurement

- ☆ A diameter is a straight line passing through the center of a circle or sphere and meeting at each end of circumference or surface.
- ☆ The most common diameter measurements taken in forestry are of the main stem of standing trees, cut portions of trees and branches.
- ☆ Diameter measurement is important because it is one of the directly measurable dimensions from which tree cross sectional area and volume etc. can be computed.
- ☆ The point at which diameters are measured will vary with circumstances.

DBH Measurement and its Significance

- ☆ The most frequent tree measurement made by forester is diameter at breast height (dbh).
- ☆ DBH is defined as the average stem diameter outside bark, at a point 1.3 m above ground as measured
- ☆ The rational of DBH measurement of individual trees is to estimate the quantity of timber, fuel wood or any other forest products which can be obtained from them.
- ☆ These measurements are also necessary for making inventory of growing stock as well as to correlate height, volume, age, increment with most easily determinable dimension i.e. dbh

DBH has been accepted as the standard height for diameter measurement because ...

- It is a convenient height for taking measurement.
- It is economic (the base of the tree is generally covered with the grasses and shrubs and even thorns sometimes).
- Majority of the trees develop root swell near the base (abnormalities at the base).
- It gives a uniform point of measurement and standardization is maintained.

Rules of DBH Measurement and Instruments Used

Rules of DBH Measurement

- Moss, creepers, lichens and loose bark found on the tree must be removed before measuring the diameter over bark.
- Breast height (BH) should be by means of a measuring stick on standing trees at 1.3m above the ground level.

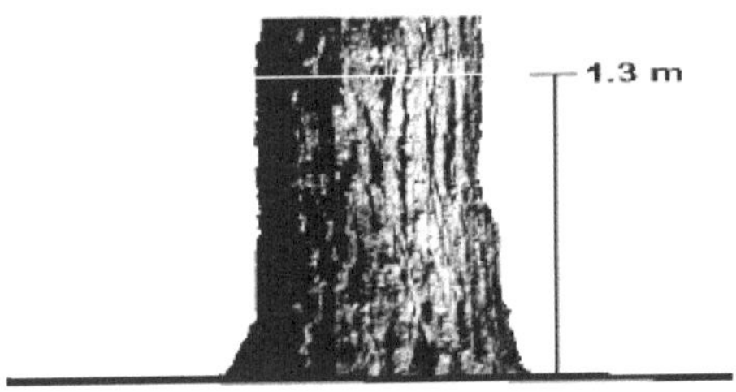

Fig. 3: Level ground

- BH point should be marked by intersecting vertical and horizontal lines 12 cm long, painted with white paint.
- On sloping land, the diameter at BH should be measured on the uphill side.

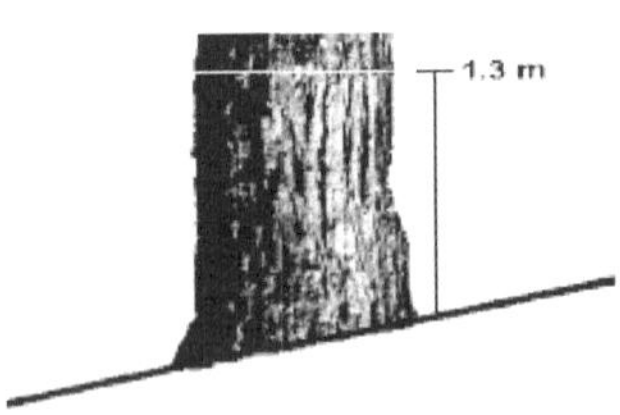

Fig. 4: Sloped ground

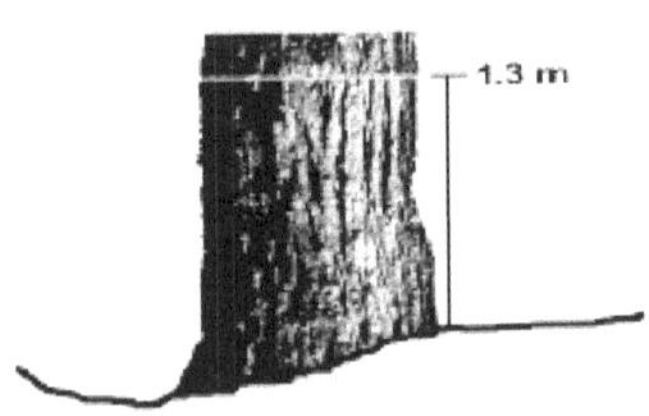

Fig. 5: Uneven ground

- ☆ In case of the tree is leaning, dbh is measured along the tree stem and not vertically, on the side of the lean for trees growing on flat ground and on the uphill side, for trees growing on sloping ground.

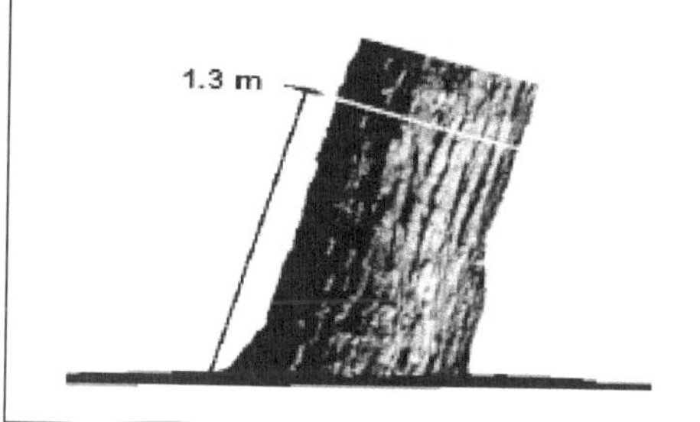

Fig. 6: Leaning Tree

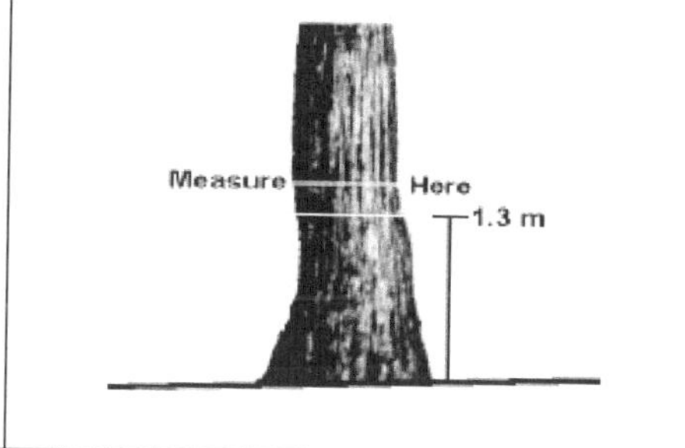

Fig. 7: Crooked Tree

- ☆ The dbh should not be measured at 1.3m if the stem is abnormal at the level. BH mark should be shifted up or down as little as possible to a more normal position of the stem and then dia. Measured.

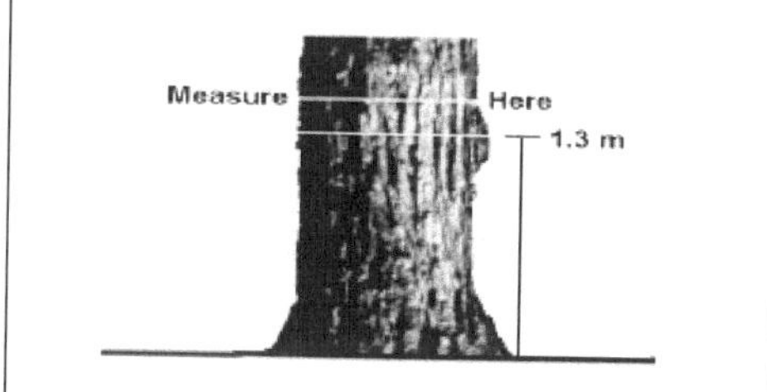

Fig. 8: Defect at 1.3m

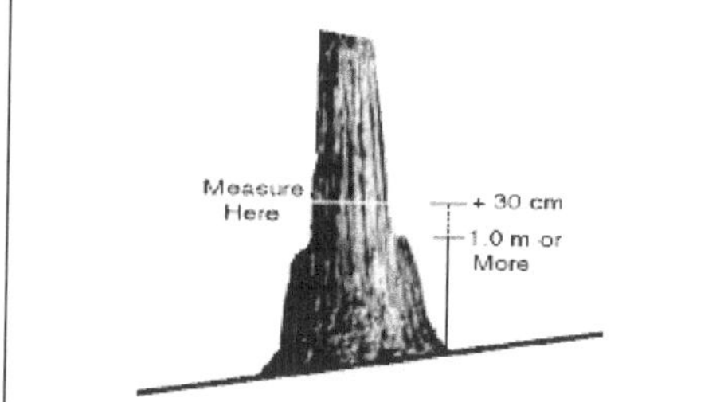

Fig. 9: Buttressed tree

- ☆ BH should be taken at the lowest point above which the buttress formation is not likely to extend
- ☆ When the tree is forked above the BH, it is counted as one tree, but when it is forked below BH, each fork should be treated as though it were a separate tree.

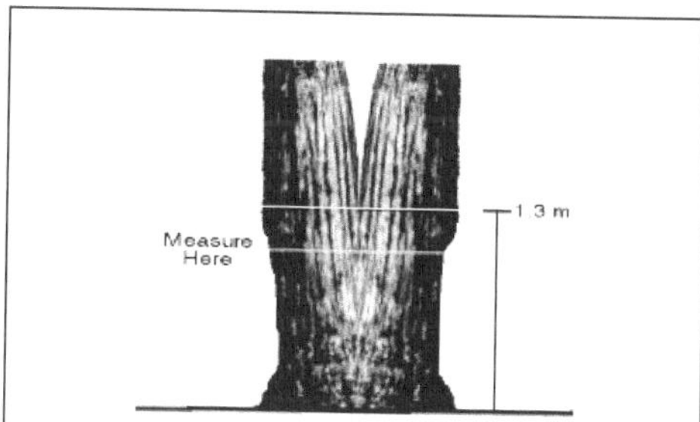

Fig. 10: Fork at 1.3 m (1 tree)

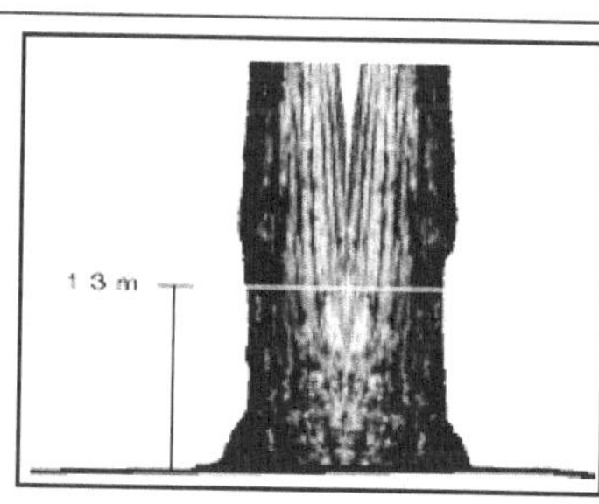

Fig. 11: Fork above 1.2 m (1 tree)

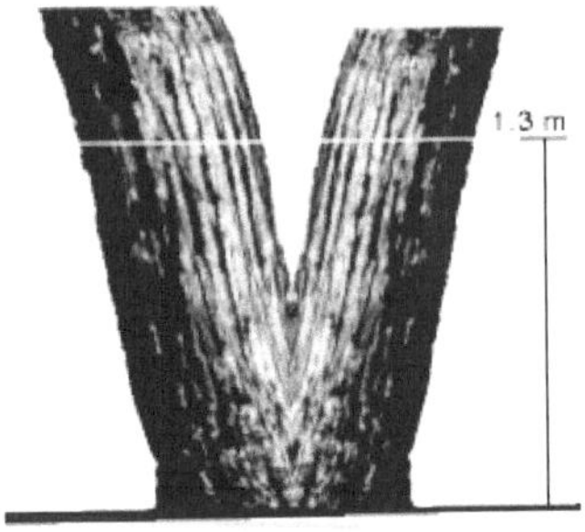

Fig. 12: Fork below 1.3 m (2 trees)

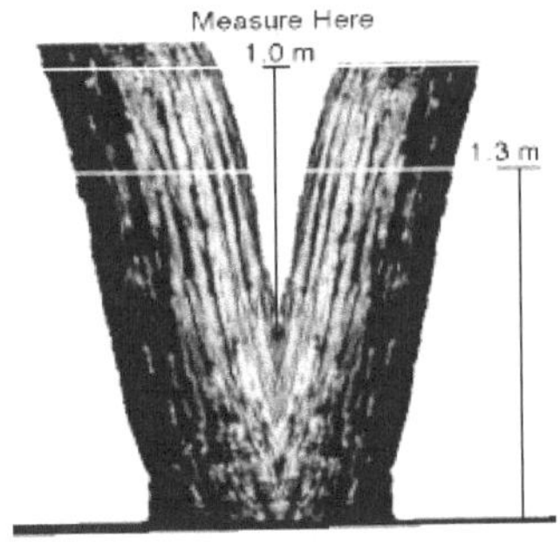

Fig. 13: Fork below 1.3m (2 trees: alternative method)

Diameter Measuring Instruments

- ☆ The most commonly used instruments for measuring diameters at BH are: Diameter tape, calipers, Biltmore stick and other optical instruments.
- ☆ Collectively, instruments employed in determining tree diameters are referred to as dendrometers.

Biltmore Stick

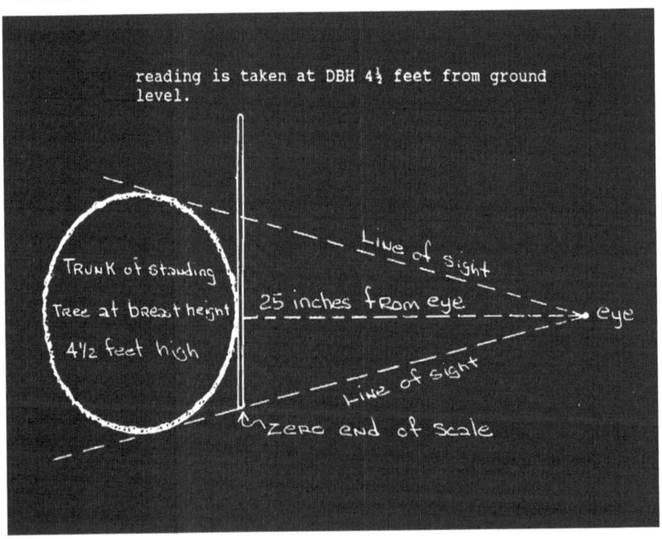

- ☆ It is a specially graduated stick used for diameter measurement.
- ☆ The stick measures a tangent to a circle, from a point, exactly 25 inches from the tree. (The stick is graduated to show the diameter as if it were projected from the user's eye into the tree.)

- ☆ This specially graduated stick is placed against the tree trunk with the diameter scale facing the user. Keep your head 25 inches from the stick, and without moving your head, slide the stick so that the left edge of the stick appears to line up exactly with the left edge of the tree trunk.
- ☆ Keep your head stationary, and move your eyes to read the number on the scale those lines up exactly with the right edge of the tree trunk. That number is a reasonable estimate of the diameter of the tree at breast height (DBH).
- ☆ If the measurement of the right edge of the tree falls between two numbers on the scale, the lower number is used as the diameter.
- ☆ The Biltmore stick tends to be inaccurate on large timber. Trees over 20 inches should be checked with a diameter tape.

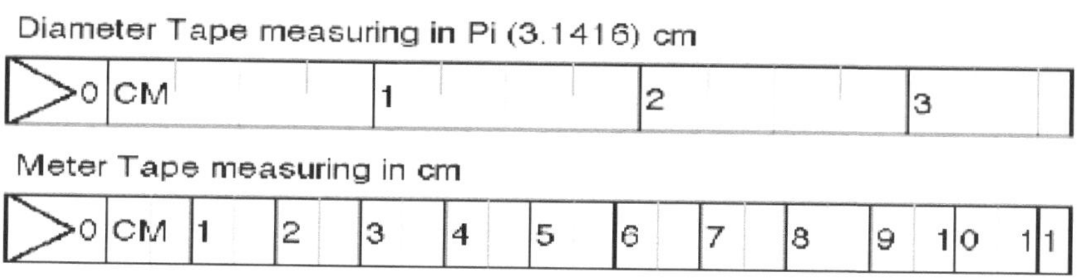

Fig. 15: Tape

Measuring Tape

- ☆ The diameter of a tree cross section may be obtained with a flexible tape by measuring the circumference of the tree and dividing by $\pi (D=C/\pi)$.
- ☆ The diameter tapes used by foresters, however are graduated at intervals of π units (in or cm), thus permitting a direct reading of diameter.
- ☆ A diameter tape is a measuring tape that has scales on both sides: one side is specially marked to show the diameter of a tree, and the other is a normal scale.

Precautions in Using Tape

- ☆ The tape should not be old.
- ☆ It must lie flat against the tree and not in twisted manner.
- ☆ It must lie in a perpendicular to the axis of the tree.

☆ The tape should be taken care of.

Advantage of Tape

☆ Tape is convenient to carry.

☆ It does not require constant adjustment.

☆ Only one measurement is needed even with irregular trees.

☆ Diameter measurement by tape is the easiest in the case of logs lying on ground.

☆ The errors in case of tape are always positive and systematic.

☆ Tape negotiates the whole circumference of the tree.

☆ Tape readings are more consistent.

Disadvantages of Tape

☆ The tape exaggerates the diameter if the tree has rough bark.

☆ It is somewhat slower to particularly in areas with dense shrub growth.

☆ Difference in tension of the tape due to elasticity affects true diameter.

Calipers

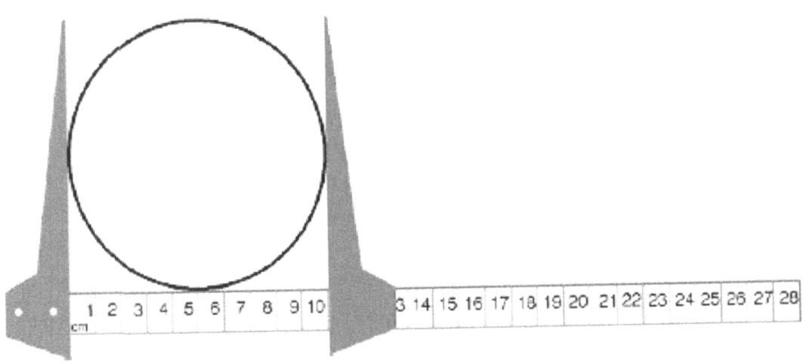

Fig. 16: Caliper

Calipers are often used to measure tree dbh or when diameters are less than about 60 cm.

☆ A caliper may be constructed of metal, plastic or wood, consists of a graduated beam/rule with two perpendicular arms.

☆ One arm is fixed at the origin of the scale and the other arm slides. When the beam is pressed against the tree and the arms closed, the beam of the caliper can be read on the scale.

- For an accurate reading, the beam of the caliper must be pressed against the tree with the beam perpendicular to the axis of the tree stem and the arms parallel and perpendicular to the beam.

Diameter Measurement Using Calipers

- Place the calipers over the stem at the required height.
- Record the diameter then take another measurement at a right angle to the first and record this measurement and
- Calculate the average of the two measurements and record to the nearest to 0.1cm.

Precautions in Use

- The calipers must be placed on the tree with movable arm well opened and must not be forced on the tree.
- The reading must be taken before the caliper is removed from tree.
- If the cross section of the stem is more or less elliptical, it is necessary to measure two diameters.
- Calipers must be placed at right angles to the axis of the tree.
- The two arms of the caliper must be in contact with the tree and the movable arm should be at right angles to the scale arm.
- Not only should the two arms of the caliper be in contact with the tree but the scale arm must also touch it.

Advantages

- Diameters can be read directly in centimeters and millimeters, thus making the instrument applicable for precise scientific work.
- By pressing the arms against the tree bole, the loose swollen bark is crushed out and irregularity from this source is avoided.
- It is adaptable for use by unskilled labour.
- The errors are both positive and negative and therefore the chances are that they may neutralize to give more accurate results than the tape.

Disadvantages

- They are not accurate when not in adjustment.
- Calipers sufficient in size to measure large trees are very awkward to carry and handle.

- Two measurements have to be taken on every tree to get the correct diameter.
- Movable arms often stick when the scale is wet or dirty, thus wasting a lot of time.

Height Measurement

- Height is the linear distance of an object normal to the surface of the earth.
- Tree height is the vertical distance measured from the ground surface.
- Height of standing tree is measured to find out its volume. Heights of selected trees in a forest are also required to read volume tables, form factor tables, yield tables etc.
- Lastly, heights of trees are required to find out productive capacity of site. Height is generally considered as an index of fertility and with the knowledge of age it gives a reliable measure of the site quality of a locality.

Total height of a standing tree is the distance along the axis of the tree stem between the ground and the tip of the tree.

Bole height is the distance along the axis of tree between ground level and Crown Point. (Crown Point is the position of the first crown forming branch).

Commercial bole height is the height of bole that is usually fit for utilization as timber.

Height of standard timber bole is the height of the bole from the ground level up to the point where average diameter over bark is 20cm.

Stump height is the distance between the ground and basal position on the main stem where a tree is cut.

Fig. 17: Tree Height

Crown length-The vertical measurement of the crown of the tree from the tip to the point half way between the lowest green branches forming green crown all round and the lowest green branch on the bole.

Crown height - The height of the crown as a measured vertically from the ground level to the point half way between the lowest green the lowest green branches forming green crown all round.

Principles of Height Measurement

- ☆ Instruments used for measuring tree heights are collectively referred to as hypsometers.
- ☆ All height measuring instruments are based either on geometric principles of similar triangles or on trigonometric principles based on relations between the sides of right-angled triangle.

Trigonometric Principles

The principles follow the basic rules of trigonometry for deriving heights of trees from distance and angle measurements. Two laws are applicable for this purpose and they are: tangent law and sine law. Instruments based on Trigonometric principles are Brandis hypsometer, Abney's level, Haga Altimeter, Topographical Abney's level, Relaskop, Tele Relaskop, Barr and Stroud dendrometer, Blume-Leiss hypsometer.

Tangent Law

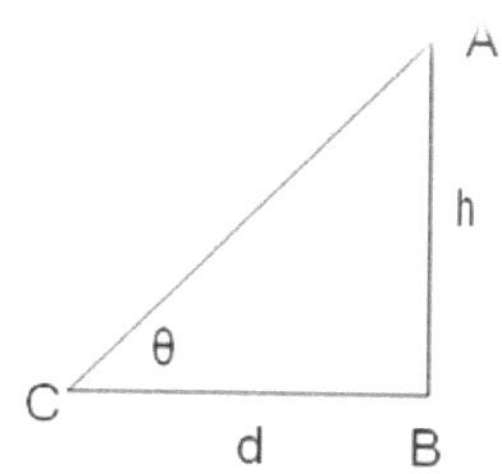

- ☆ Applicable to right angle triangle
- ☆ For accurate results, trees must not lean more than 5° from the vertical, and the fixed horizontal distance must be determined by taped measurement.

$$\tan\theta = AB/BC,\ h = d \times \tan\theta$$

Sine Law

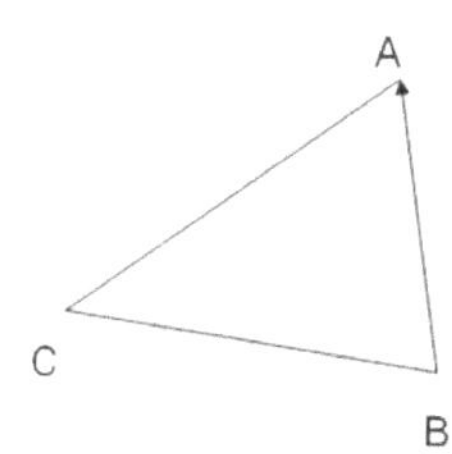

- ☆ Applicable to non right angle triangle is useful in deriving tree height in difficult conditions.
- ☆ Sines of angles are proportional to the opposite sides.

Sin <ACB/AB = sin < CAB/BC = sin < ABC/AC

Geometric Principle of Similar Triangle

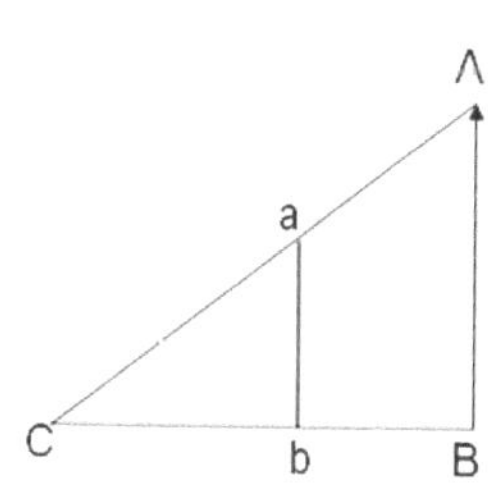

- ☆ Corresponding angles are equal and the corresponding sides are proportional.
- ☆ By knowing the two sides of a triangle and only one side of the other, the corresponding second side of the latter can be found.
- ☆ Useful in rough estimation, not reliable for precise work. Eg. Christen's hypsometer, Smythies Hypsometer etc.

AB/ab = BC/bC,

AB = ab × BC/bC

Non Instrumental Methods

- **Shadow method**: A pole of convenient length is fixed upright in the ground and its height above the ground is measured. The shadows of the pole and the tree are also measured.

  AB/ab = BD/bd, AB = BD × ab/bd

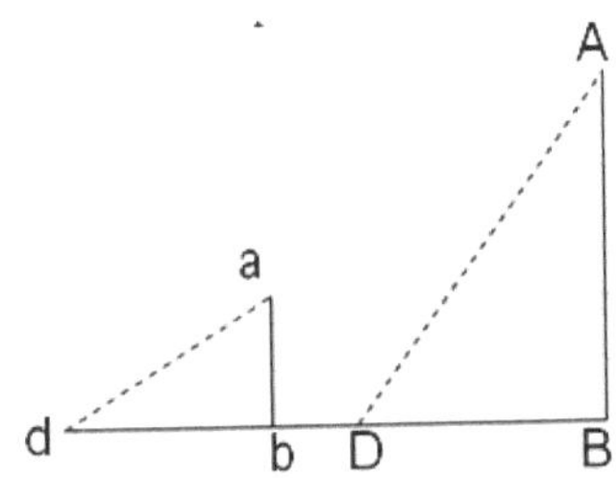

Where, AB is the tree, ab is the portion of the pole above the ground level, BD is the length of shadow of the tree and bd is the shadow of ab.

- **Single pole method**: Pole of about 1.5 m length vertically at arm's length in one hand in such a way that portion of the pole above the hand is equal in length to the distance of the pole from eye.

  AB/ab = EB/Eb i.e. AB = EB x ab/Eb

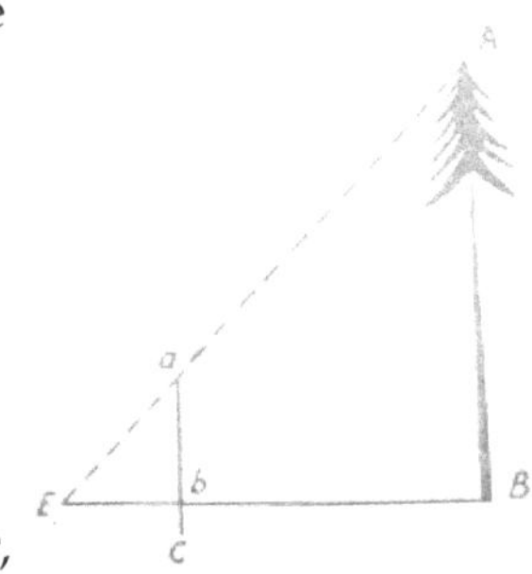

Where,

AB = tree, ac=pole about 1.5 m long, Eb=ab

Instrumental Method

- By using instruments like hypsometer, clinometer, altimeters, abney's level etc.
- All these instruments are based either on geometric principle of similar triangles or on trigonometric principles.

Measurement of Height (vertical & leaning) Tree in Plane and Slope Areas

Measurement of Height of Trees on Plane Area

- The height of the tree is calculated with the help of the tangents of the angle to the top and the distance of the observer from the tree.

  AB = AD + BD = ED tan α + BD = BF tan α + EF

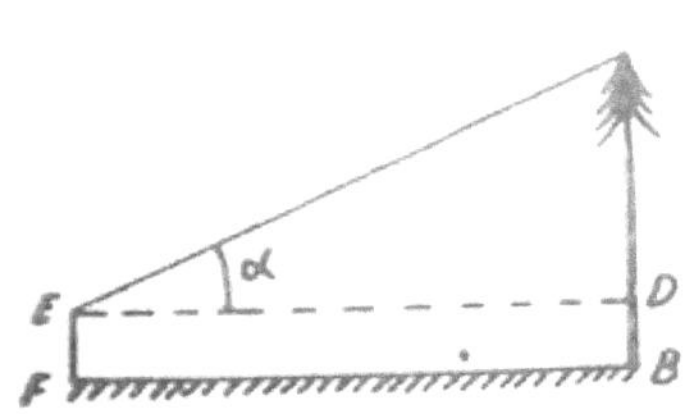

Where, AB = tree, EF = eye height of the observer,

BF = horizontal distance

Measurement of Height of Trees on Sloped Area

☆ Where the observer is standing at such a place that the top of the tree is above the eye level and the base below it.

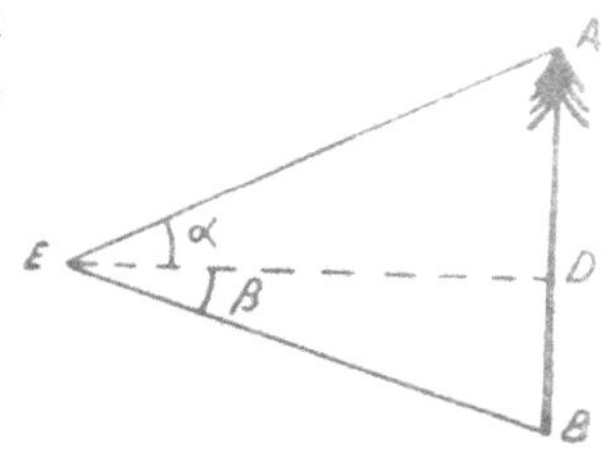

AB = AD + DB

= ED tan α + ED tan β = ED (tan α + tanβ)

= EB Cosβ (tan α + tanβ)

☆ Where top and base of the tree are above the eye level.

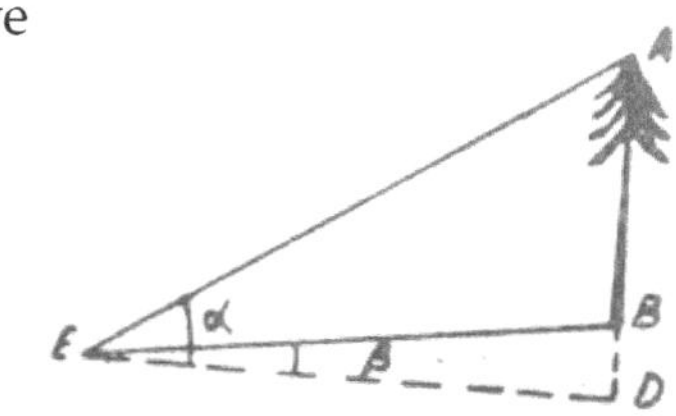

AB = AD-BD

= ED tan α – ED tan β

= ED (tanα-tan β)

= EB cos β (tanα-tan β)

☆ Where base and top of the tree are below the eye level

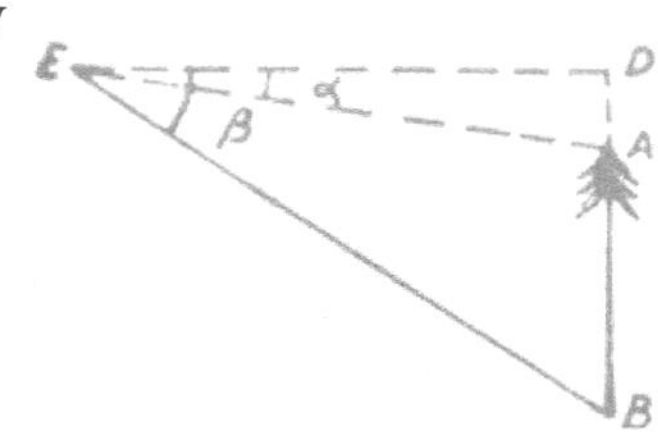

AB = BD – AD

= ED tan β – ED tan α

= ED (tan β - tan α)

= EB cos β (tan β - tan α)

Measurement of Height of Leaning Tree

Condition-1(a): In case of the observer standing at between the top and bottom of the tree (lean away from the observer)

exterior $\angle$ECB = interior $\angle$ACB + $\angle$CBA

Therefore

$\angle$CAB = $\angle$ECB - $\angle$CBA

=(90 °- α)- θ

90 ° - (α+ θ)

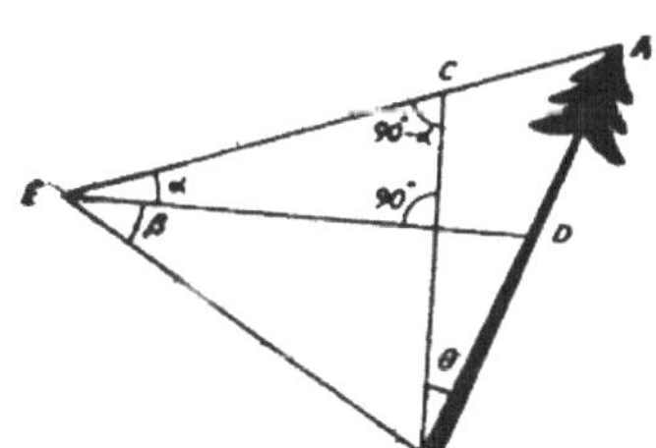

Now in triangle AEB,

$$\frac{AB}{Sin\angle AEB} = \frac{EB}{Sin\angle EAB}$$

Therefore,

$$AB = \frac{EBSin\angle AEB}{Sin\angle EAB}$$

$$= \frac{EBSin(\alpha+\beta)}{Sin[90^0-(\alpha+\theta)]}$$

$$= \frac{EBSin(\alpha+\beta)}{Cos(\alpha+\theta)}$$

Condition-1(b): In case of the observer standing at between the top and bottom of the tree (lean towards the observer)

In the triangle ACB,

exterior ∠EAB = interior ∠ACB + ∠ABC

$=90°-\alpha+\theta$

$= 90°-(\alpha-\theta)$

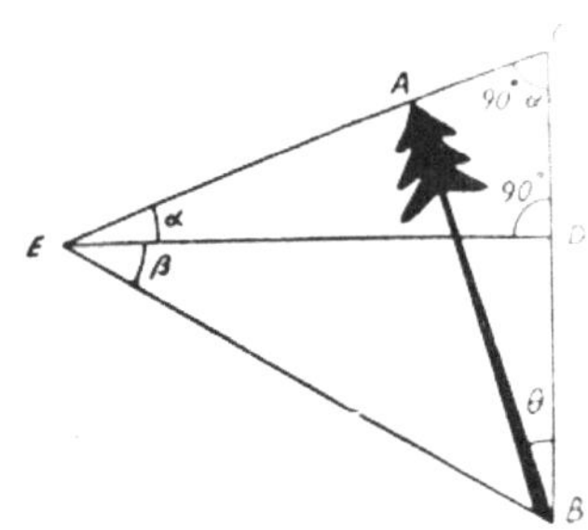

Now in triangle AEB,

$$\frac{AB}{Sin\angle AEB} = \frac{B}{Sin\angle EAB}$$

Therefore,

$$AB = \frac{EBSin\angle AEB}{Sin\angle EAB}$$

$$= \frac{EBSin(\alpha+\beta)}{Sin[90^0-(\alpha-\theta)]}$$

$$= \frac{EBSin(\alpha+\beta)}{Cos(\alpha-\theta)}$$

Condition-2(a): When the observer is below the top and bottom of the tree (lean away from the observer)

In the triangle ACB,

∠CAB = exterior ∠ECB - interior ∠CBA

$= 90°-\alpha-\theta$

$= 90°-(\alpha+\theta)$

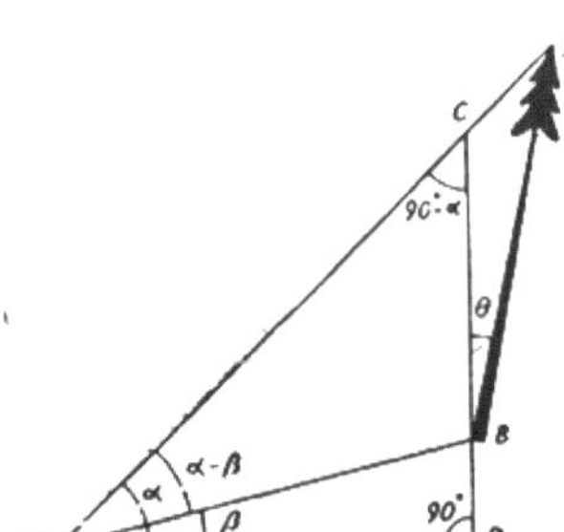

Now in triangle AEB,

$$\frac{AB}{Sin\angle AEB}=\frac{EB}{Sin\angle EAB}$$

Therefore,

$$AB=\frac{EBSin\angle AEB}{Sin\angle EAB}$$

$$=\frac{EBSin(\alpha-\beta)}{Sin[90^0-(\alpha+\theta)]}$$

$$=\frac{EBSin(\alpha-\beta)}{Cos(\alpha+\theta)}$$

Condition-2(b): when the observer is below the top and bottom of the tree (lean towards the observer)

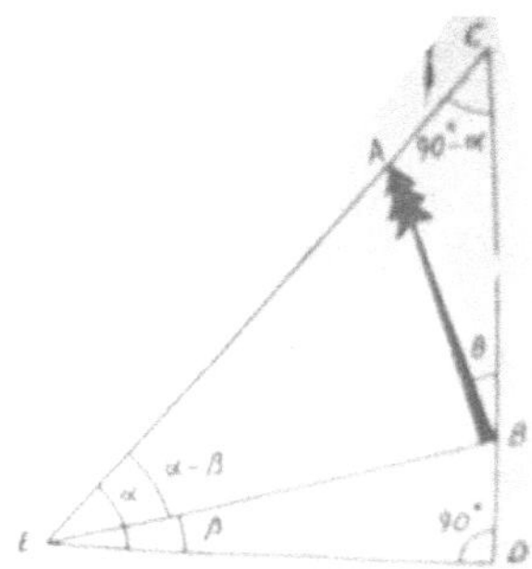

In the triangle ACB,

exterior ∠EAB = interior ∠ACB + ∠ABC

= 90°-α + θ

= 90°-(α - θ)

Now in triangle AEB,

$$\frac{AB}{Sin\angle AEB}=\frac{EB}{Sin\angle EAB}$$

Therefore,

$$AB=\frac{EBSin\angle AEB}{Sin\angle EAB}$$

$$=\frac{EBSin(\alpha-\beta)}{Sin[90^0-(\alpha-\theta)]}$$

$$=\frac{EBSin(\alpha-\beta)}{Cos(\alpha-\theta)}$$

Condition-3(a): when the observer is above the top and bottom of the tree (lean away from the observer)

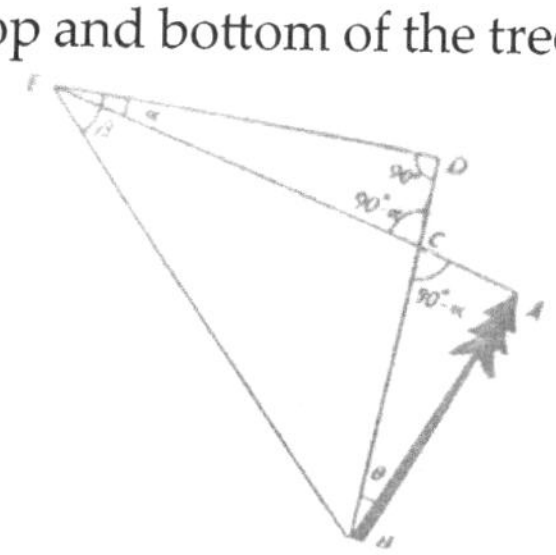

In the triangle ACB,

∠CAB = 180°-[∠ACB+ ∠ABC]

= 180°-[90°-α + θ

= 90°+(α - θ)

Now in triangle AEB,

$$\frac{AB}{Sin\angle AEB} = \frac{EB}{Sin\angle EAB}$$

Therefore,

$$AB = \frac{EBSin\angle AEB}{Sin\angle EAB}$$

$$= \frac{EBSin(\beta - \alpha)}{Sin[90^0 + (\alpha - \theta)]}$$

$$= \frac{EBSin(\beta - \alpha)}{Cos(\alpha - \theta)}$$

Condition-3(b): when the observer is above the top and bottom of the tree (lean is towards the observer)

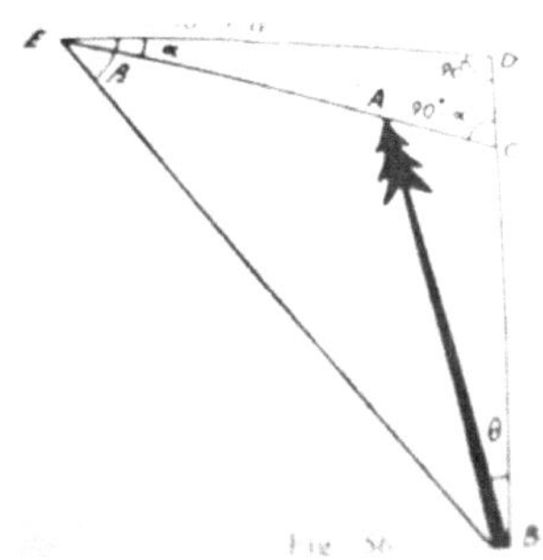

In the triangle ACB,

exterior ∠EAB = interior ∠ACB + ∠ABC

$= 90° + \alpha + \theta$

$= 90° + +(\alpha + \theta)$

Now in triangle AEB,

$$\frac{AB}{Sin\angle AEB} = \frac{EB}{Sin\angle EAB}$$

Therefore,

$$AB = \frac{EBSin\angle AEB}{Sin\angle EAB}$$

$$= \frac{EBSin(\beta - \alpha)}{Sin[90^0 + (\alpha + \theta)]}$$

$$= \frac{EBSin(\beta - \alpha)}{Cos(\alpha + \theta)}$$

Instrument Used in Height Measurement

- ☆ There are various instruments to measure height of the tree.
- ☆ Height measuring instruments are called hypsometer.
- ☆ Those instruments based on trigonometric principles are more accurate than the ones employing geometric principles.
- ☆ The Abney's level, Haga Altimeter, Blume-Leiss Altimeter and Sunto Clinometer are similar in accuracy.

a. Christen's Hypsometer

- ☆ It is based on the geometric relationships of similar triangles.
- ☆ Consists of a strip of metal, thin wood or card board about 2.5cm wide and 33 cm length.
- ☆ It has two flanges or protruding edges one at the top and other at the bottom.
- ☆ Each flange has a hole in it, the upper one to suspend the instrument by some thread passing through it at and the lower one to suspend a weight from it to prevent it from swinging.

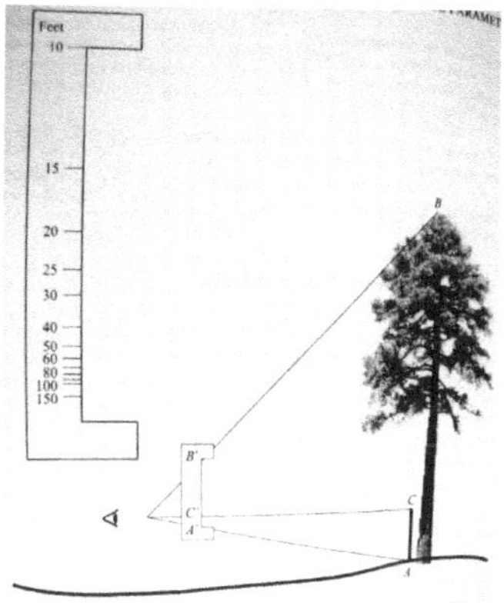

Fig. 18: Christen hypsometer

- ☆ To use it, a pole (usually 5 or 10 ft long) is held upright against the base of the tree, or a mark is placed on the tree at a height of 5 or 10 ft above the ground.
- ☆ The hypsometer is then held vertically at a distance from the eye such that the two inside edges of the flanges are in line with the top and base of the tree.
- ☆ It may be necessary for the observer to move closer to or farther from the tree to accomplish this, but except for this, the distance from the tree is immaterial.
- ☆ The graduation on the scale that is in line with the top of the pole, or the mark, gives the height of the tree.
- ☆ The following proportion gives the formula for graduating the instrument.

$$\frac{A'C'}{AC} = \frac{A'B'}{AB}$$

$$AB = \frac{AC * A'B'}{A'C'}$$

For a given length of instrument A'B' and a given pole length or mark height AC, the graduation A'C' can be obtained by substituting different values of height AB in the equation.

- ☆ Although the christen hypsometer may be used to measure any type of height, it is practical only for total height measurements.
- ☆ A crowding of graduations at the bottom of the scale, makes the instrument unreliable for the determination of the height of tall trees.

Advantages

- ☆ It is light, easily made and easy to transport
- ☆ The height of the tree can be read directly.
- ☆ It is quicker to use and so it is useful in conditions where speed is required.

Disadvantages

- ☆ Extra care has to be taken to hold the top and bottom of the tree within the flanges while reading the heights.
- ☆ It should be held in the true vertical plane
- ☆ It is not suitable for more than 30m tree height.
- ☆ It requires the use of staff.
- ☆ Skill is necessary to use the instrument with consistent accuracy.

b. Sunto Clinometers

- ☆ Hypsometers based on tangent of angles such as Abney's level, Haga altimeter, the Blume-Leiss altimeter and the Sunto clinometers are used in height measurement of trees.
- ☆ The sunto clinometer is a handheld device house in a corrosion-resistant aluminum body.

Fig. 19: Sunto Clinometer

- ☆ A jewel-bearing assembly supports the scale, and all moving parts are immersed in a damping liquid inside a hermetically sealed plastic capsule.
- ☆ The liquid dampens undue scale vibrations.
- ☆ The instrument is held to one eye and raised as lowered until the baseline is seen at the point of measurement.
- ☆ At the same time, the position of the hairline on the scale gives the reading.
- ☆ Due to optical illusion, the hairline seems to continue outside the frame and can be observed at the point of measurement.

- ☆ The instrument is available with several scale combinations: percent and degrees, percent and topographic, degrees and topographic, and feet and meter.
- ☆ Hypsometers based on the tangents of angle are more accurate than those on similar triangles.
- ☆ When used correctly, the Suunto Clinometer has an accuracy of about +/- 0.5 m for a 20 m tall tree (ie about 2.5%).

Use

- ☆ Measure the horizontal distance from the base of a vertical tree (or the position directly beneath the tree tip of a leaning tree) to a location where the required point on the tree (e.g. tree tip) can be seen.
- ☆ Sight at the required point on the tree:
  - **Using one eye**: Close one eye and simultaneously look through the Suunto at the scale and 'beside' the Suunto at the tree. Judge where the horizontal line on the Suunto scale would cross the tree.
  - **Both eyes**: With one eye looking at the Suunto scale and the other looking at the tree, allow the images to appear to be superimposed on each other and read where the horizontal line on the Suunto scale crosses the tree. Note: If you suffer from astigmatism (a common situation where the eyes are not exactly parallel), use the one eye approach.
- ☆ Read from the percent scale and multiply this percentage by the horizontal distance measured in step 1.
- ☆ Site to the base of the tree and repeat steps 2 - 3.
- ☆ Combine the heights from steps 3 and 4 to determine total tree height:
  - Add the 2 heights together if you looked up to the required point in step 2 and down to the base of the tree in step 4.
  - Subtract the height to the base of the tree from the height to the required point if you are on sloping ground and had to look up to both the required point and the base of the tree.
- ☆ Check all readings and calculations.

c. Abney's level

- ☆ It is used to measure tree heights as well as land elevations.
- ☆ The instrument consists of a graduated arc mounted on a sighting tube about 6 inches long.

- ☆ The arc may have a degree, percentage or topographic scale.
- ☆ When the level bubble, which is attached to the instrument, is rotated while a sight is taken, a small mirror inside the tube makes it possible to observe when the bubble is horizontal.

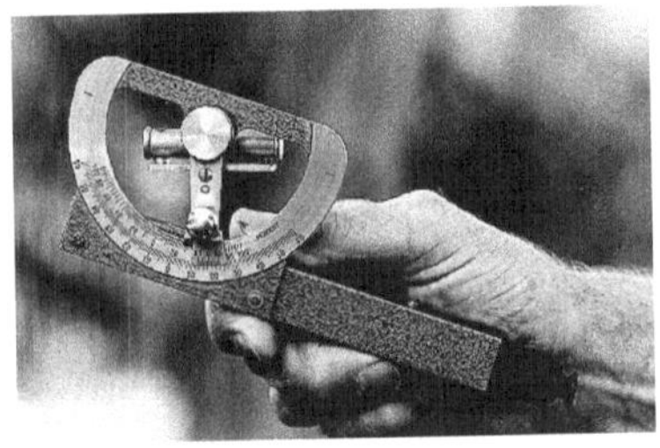

Fig. 20: Abney's level

- ☆ The angle between the bubble tube and the sighting tube may be read on the arc.
- ☆ The abney's level, however, is slower to use, and large vertical angles are difficult to measure because of the effect of refraction on observations of the bubble through the tube beneath.
- ☆ This makes the abney level difficult to use in tall timber that is so dense that the tops cannot be seen from a considerable distance.
- ☆ When used correctly, the Abney Level has an accuracy of about +/- 0.5 m for a 20 m tall tree (ie about 2.5%).

Use

- ☆ Measure the horizontal distance from the base of a vertical tree (or the position directly beneath the tree tip of a leaning tree) to a location where the required point on the tree (e.g. tree tip) can be seen.
- ☆ Sight at the required point and move the index arm over the scale until the bubble tube is level.
- ☆ Read the percentage scale (or the degrees and minutes of the angle).
- ☆ Calculate the height by multiplying the percentage read by the horizontal distance (or by multiplying the horizontal distance by *Tan* of the angle).
- ☆ Site to the base of the tree and repeat steps 2 - 4.
- ☆ Combine the heights from steps 4 and 5 to determine total tree height:
  - Add the 2 heights together if you looked up to the required point in step 2 and down to the base of the tree in step 5.
  - Subtract the height to the base of the tree from the height to the required point if you are on sloping ground and had to look up to both the required point and the base of the tree.
- ☆ Check all readings and calculations.

Advantages

- ☆ It gives accurate angles of elevation and depression
- ☆ Reading can be taken after sighting the tree without disturbing the index.
- ☆ It is small and light and can be used even in hills without difficulty.

Disadvantages

- ☆ Shaking of the hand makes the sighting of the top or bottom of the tree a little difficult and time consuming.
- ☆ The spirit level has to be adjusted by moving the head of the screw while simultaneously looking to the top or bottom of the tree.

d. Haga Altimeter

It consists of a gravity-controlled, damped, pivoted pointer, and a series of scales on a rotatable, hexagonal bar in a metal, pistol-shaped case.

Fig. 21: Haga altimeter

- ☆ The six regular American scales are 15, 20, 25, 30, percentage, and topographic scale.
- ☆ Sights are taken through a gun-type peep sight; squeezing a trigger locks the indicator needle, and the observed reading is taken on the scale.
- ☆ A range finder is available with this instrument.
- ☆ When used correctly, the Blume Leiss has an accuracy of about +/- 0.5 m for a 20 m tall tree (ie about 2.5%).

Use

- ☆ Select a location, preferably 15, 20, 30 or 40 meters horizontal distance from the base of a vertical tree (or the position directly beneath the tree tip of a leaning tree) where the required point on the tree (e.g. tree tip) can be seen.
- ☆ Select the appropriate distance scale on the rotating rod.
- ☆ Release the pointer by pressing the button on the side of the instrument.
- ☆ Sight at the required point on the tree, wait for a moment for the pointer to settle then pull trigger.

☆ Read the height directly from the appropriate scale if you are 15, 20, 30, or 40 meters away from the tree. If you were unable to find a position at one of these distances:
  - If the horizontal distance is a simple fraction of one of the scale distances (e.g. 10 m is half of 20 m), read from the scale distance and multiply by the appropriate fraction.
  - Read from the percent scale and multiply this percentage by the horizontal distance measured in step 1.

☆ Site to the base of the tree and repeat steps 3 - 5.

☆ Combine the heights from steps 5 and 6 to determine total tree height:
  - Add the 2 heights together if you looked up to the required point in step 2 and down to the base of the tree in step 6.
  - Subtract the height to the base of the tree from the height to the required point if you are on sloping ground and had to look up to both the required point and the base of the tree.

☆ Check all readings and calculations.

e. Spiegel Relaskop

The Spiegel Relaskop, also known as a Relaskop, is a sophisticated instrument that can be used to measure stand basal area and tree height and diameter at any point up a tree bole.

Fig. 22: Spiegel relaskop

☆ In conjunction with other equipment, the Relaskop can be used in the estimation of distance (range) to an object and the number of trees / ha.

☆ The Relaskop has a peep-hole at the rear and a clear window at the front.

☆ Three additional windows in the lower half of the instrument allow light to enter and illuminate the scale.

☆ A brake button, bottom half at the front of the instrument, allows a weighted wheel within the Relaskop to rotate.

☆ When looking through the peephole, a circular field of view is seen.

☆ The scales are seen in the bottom half of this field of view and scale readings are taken where the scale touches the line halfway up the field of view.

- ☆ The standard metric Relaskop has three scales for measuring (vertical) height. The appropriate scale will depend on the horizontal distance from the tree.
  - left-most scale - 20 m from the tree.
  - middle-left - 25 m from the tree.
  - middle-right - 30 m from the tree.
- ☆ If you depress the brake button and look straight up or down, the appropriate distance values can be seen alongside their scales.
- ☆ Select a point from where base and tip (or any other points of interest) must be clearly visible from the selected point.

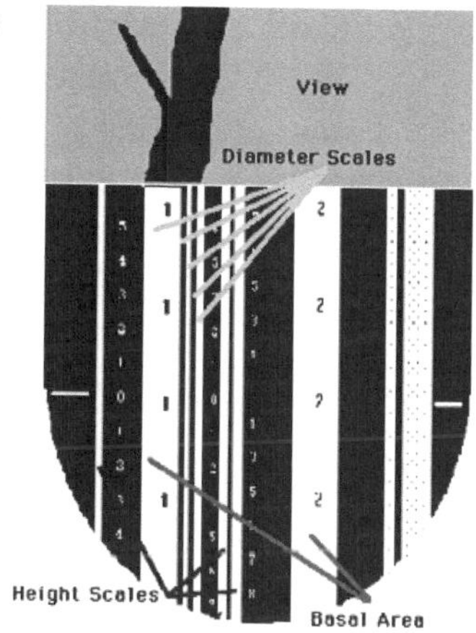

Fig. 23: Scales in Spiegel Relaskop

f. Blume Leiss Altimeter

It is similar in construction and operation to the Haga altimeter, although its appearance is somewhat different.

- ☆ The regular scales are 15, 20, 30 and 40. A degree scale is also provided.
- ☆ All scales can be seen at a time. The instrument is available with a rangefinder.

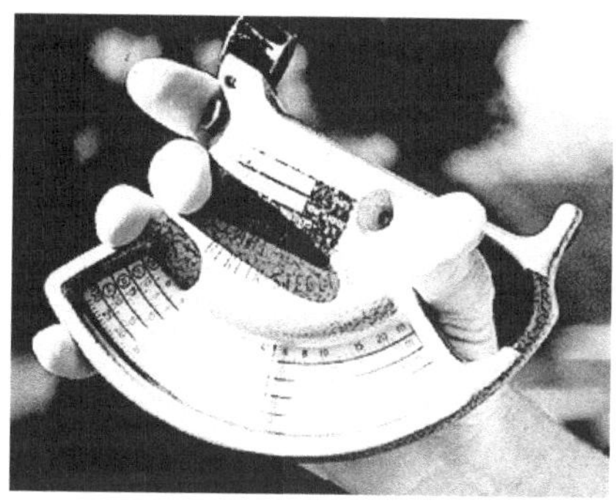

Fig. 24: Blume leiss altimeter

- ☆ When used correctly, the Blume Leiss has an accuracy of about +/- 0.5 m for a 20 m tall tree (ie about 2.5%).

Use

- ☆ Select a location, preferably 15, 20, 30 or 40 meters horizontal distance from the base of a vertical tree (or the position directly beneath the tree tip of a leaning tree) where the required point on the tree (e.g. tree tip) can be seen.
- ☆ Release the pointer by pressing the button on the side of the instrument.
- ☆ Sight at the required point on the tree, wait for a moment for the pointer to settle then pull trigger.

- ☆ Read the height directly from the appropriate scale if you are 15, 20, 30, or 40 meters away from the tree. If you were unable to find a position at one of these distances:
  - If the horizontal distance is a simple fraction of one of the scale distances (e.g. 10 m is half of 20 m), read from the scale distance and multiply by the appropriate fraction.
  - Read from the percent scale and multiply this percentage by the horizontal distance measured in step 1.
- ☆ Site to the base of the tree and repeat steps 2 - 4.
- ☆ Combine the heights from steps 4 and 5 to determine total tree height:
  - Add the 2 heights together if you looked up to the required point in step 2 and down to the base of the tree in step 5.
  - Subtract the height to the base of the tree from the height to the required point if you are on sloping ground and had to look up to both the required point and the base of the tree.
- ☆ Check all readings and calculations

g. Vertex IV and Transponder T3

The vertex is primarily designed to measure the height of standing objects, and most often trees. The instrument can also be used to measure distance, horizontal distance, angle and inclination. The vertex instrument has with its ultrasonic measuring technique proved to be especially useful in dense terrains with thick undergrowth, where conventional methods such as measuring tapes, laser instruments and mechanical height measurers are difficult to use.

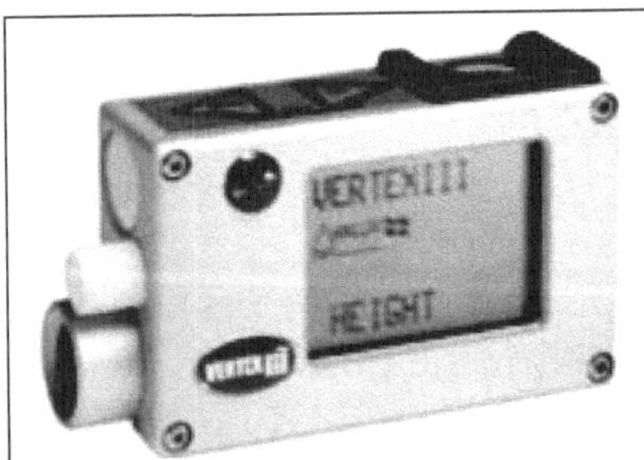

Fig. 25: Vertex

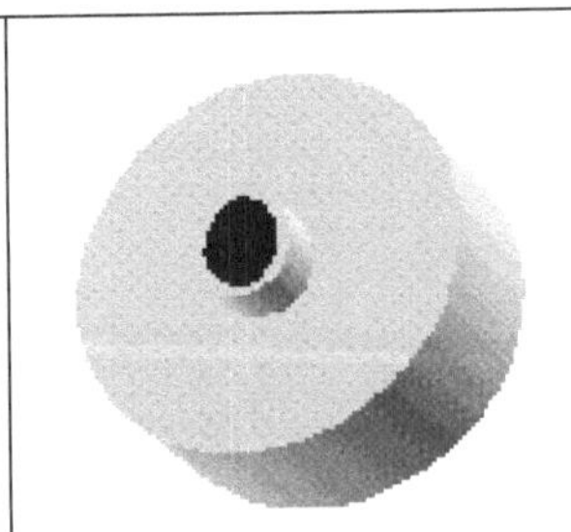

Fig. 26: Transponder 3

To define a reference point is a secure and reliable way, the vertex works with the transponder. The vertex communicates with the transponder. This communication eliminates in an efficient way any mix-ups of signals from other instruments or places (echoes). A measuring operation will not in any significant way be disturbed by objects in between the vertex and the transponder. This reference is used as a sight mark for height measuring and can be placed at optional height, where visibility is the best in for example thick vegetation. The reference point height is set in a special menu in the vertex instrument and automatically added to the measured height.

The vertex uses ultrasound to measure distances. Unlike for example measuring tapes and laser instruments, ultrasound can be used also when there is no free aim to the reference point. The ultrasound will not pass through an obstacle, but looks for the shortest way around it. Heights are calculated trigonometrically using the variables contained when measuring angle and distance. The vertex automatically assumes that the measuring object is perpendicularly positioned to the ground. With the vertex, an unlimited number of heights per object can be measured. The instrument display can show the 4 lastly measured heights per object at a time.

When using a relaskopic method to measure, an in built BAF (Basal Area Factor) can be used for the vertex instrument to control the minimum diameter for trees. The function is useful when some trees in an area are covered by other, making the decision whether to include the tree or to exclude if from the area difficult. By simply measuring the distance between the tree and the plot centre, the vertex can calculate the minimum diameter the tree should have in order to be included into the counting.

Data can be sent through IR or Bluetooth and results can be stored and processed in for example the Digitech Professional Caliper, other PC or handheld computer.

Important Facts

The Vertex uses ultra-sonic signals to determine distances. Humidity, air pressure, surrounding noise and, above all, the temperature can affect the range and extension of the ultra sonic signals. The Vertex has a built-in temperature sensor that automatically compensates for the divergence caused by variations in temperature. In some cases, distances of 50 meters and greater can be measured without problems, and in other cases, the maximum distance can be shorter than 30 meters.

To increase and optimize the measuring accuracy, calibration should be made regularly. When calibrating, it is of utmost importance that the instrument has been given enough time to stabilize at ambient temperature. If, for example, the instrument is carried in an inner pocket, it can take up to 10 minutes before it has adjusted to current outdoor temperature. The measurement inaccuracy pending on temperature is approximately 2 cm/C°.

An example: Your inner pocket holds +15 C°. Outdoor air temperature is -5 C°. The measurement result will show 10,40 m and not the correct 10,00 m.

The measuring fault can be made permanent if the instrument is calibrated before reaching the correct current temperature.

- Check your instrument daily and recalibrate if necessary.
- Do not touch the temperature sensor at the front of the instrument (the metal knob between the sight and the loudspeaker).
- Never calibrate the instrument before it has reached ambient temperature.

When measuring heights, it is important to hold the instrument as straight as possible.

The trigonometric functions calculate the height based on two (2) angles and one (1) distance. The distance can be measured manually with a tape, or automatically by using the Transponder. If using a tape, the distance has to be input in the Vertex before starting (angle-) and height measuring.

How to Use the Vertex

Height

Height measuring can be performed in different ways depending on type of surroundings and operation. Heights, distance and angle can be transmitted via InfraRed (IR) to, for example the Digitech Professional Caliper or other computer device for storage/processing with IR or Bluetooth (Vertex IV BT model only) by pressing right arrow key. For height measuring 3 last measured heights with angle and distance can be transferred.

Height Measuring with Transponder

Start the transponder and place it on/towards the object to measure. *Note that the transponder should be placed at the T. HEIGHT /(transponder height) that has been set in the settings menu.*

Walk a suitable distance from the object – for optimal results the distance equals the approximate height.

- ☆ Press ON to start the Vertex and aim at the transponder. Keep pressing ON until the crosshair sight goes out momentarily. Now release ON. The Vertex has measured the distance, the angle and the horizontal distance to the transponder.
- ☆ Aim at the height to measure with the sight cross blinking. Press ON until the cross hair disappears. The first height is locked and displayed. Repeat until all heights on the object are measured.

Inclination (ANGLE)

The Vertex is an excellent instrument to measure inclination and angles in the terrain.

☆ Press ON to start the Vertex and step with the arrow keys to ANGLE and press ON.

☆ The angle window is displayed. Aim at the point where you need to know the angle.

☆ Push and press the ON until the cross disappears. Read the obtained value in display. The angle is featured in Grads, degrees and percentage.

Note that the angle is measured from the Vertex with the crosshair sight. This implies that it is not possible to use the outside of the Vertex to measure the angle of, for example a flat table surface.

Distance Measuring (DME)

To measure the distance, press the DME key (left arrow key) when the vertex is turned off. The result, the distance between the vertex and the transponder, is presented in the vertex display.

Distance Measuring with the Adapter for 360 Degrees

With the adapter, the ultrasound is spread and it is possible to measure from any direction. This is particularly useful when working in circular sample plots, where the distance from the plot centre to objects within a defined circle should be measured.

Horizontal Distance Measuring (DME)

The vertex can be used as horizontal distance measurer (DME). The display text will rotate 90° to simplify reading the results when measuring horizontal distances.

☆ Press ON to start the vertex and go with the arrow keys to ANGLE and push ON.

☆ Aim at the point where you need to know the angle. Push and press the ON until the Red Cross goes out.

☆ To measure the distance, now push the left arrow key. The vertex presents the horizontal distance in the display.

Basal Area Factor

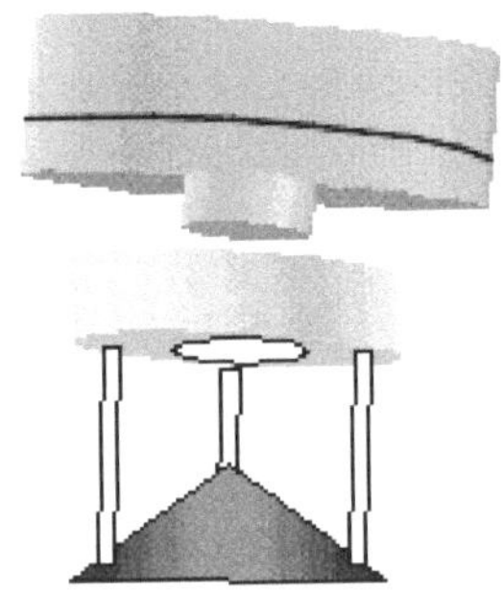

Fig. 27: Transponder with Adapter

Working in dense forest with relascopes or prism can sometimes offer difficulties and accurate diameter estimation hard to make. When using relaskopic method to measure, an in built BAF function can be used for the Vertex instrument to control the minimum diameter for trees. The function is useful when some trees in an area covered by other, making the decision whether to include the tree or to exclude if from the area difficult.

By simply measuring the distance between the tree and the plot centre, the vertex can calculate the minimum diameter the tree should have in order to be included in the counting.

Basal area factors: 0.5, 1, 2, 3, 4, 5, 6, 7, 8, 9 (m^2/ha)

Transponder

The transponder is an ultrasonic transmitter/receiver that communicates with the vertex instrument. The transponder can be used both for direct measuring (60°), and in 360° when used with "360 adapter"- for example when working in circular sample plots.

The transponder is equipped with an audible signal that tells if the transponder is activated or not.

Transponder T^3 has no switch and the vertex and /or DME is used as a remote control to turn off and on. When turned on, the transponder stays activated for app. 20 minutes.

To measure in 360° circle with the transponder, the T3 is attached to the adopter. The adapter spreads and receives the ultrasound in a full circle. The adapter can be mounted onto the custom plot center staff.

Special Considerations in Measuring Tree Heights

- ☆ It is difficult to measure accurately the height of large flat crowned trees. There is tendency to overestimate their heights.
- ☆ The optimum viewing distance for any hypsometer is the distance along the slope equal to the height to be measured. This rule of thumb should be used with discretion.
- ☆ Since all hypsometers assume that trees are vertical, tress leaning away from an observer will be underestimated and trees leaning towards

an observer will be overestimated. This error will be minimized if measurements are taken such that the lean is to the left or right of the observer.

- The measurement of tree height with an accurate hypsometer is slow and expensive.

Sources of Errors in Height Measurement

- Measuring the height of trees is time consuming and prone to errors.
- Experience has shown that when indirect methods are used to measure height, measurement from two independent positions is essential.
- The readings from the two positions should agree within the limits of instrumental error - this is an absolute check on instrument and operator error (sighted to correct tip, etc.).
- Thus, differences of up to 1 m in readings for a 40 m tree are acceptable - precision of instruments under forest conditions is no better than this.

The Most Common Errors Include

The sources of the major errors in height measurement are:

- Failure to measure correctly the horizontal distance from the observer to the tree
- If the distance from the observer to the tree is not measured horizontally, the observer will stand too near the tree, and the height will be overestimated by the direct reading on the instrument scale.
- Wind sway
- Wind causes tree tops to sway and this can be very serious hindrance in tree height measurement and cause serious errors. Accurate readings cannot be made in high winds. The errors may be reduced by averaging readings taken at the extremes of the sway towards and away from the observer.
- Leaning trees
- If the tree is leaning away from the observer, height will be under estimated and if the lean is towards the observer, height will be overestimated.
- Non linearity of the relationship of tree height and angle of sight.
- The smaller the angle of the sight the easier it is to define the highest point in the crown; but the nearer the angle of sight to 45° the smaller

is the error caused by an inaccurate reading of that angle. The best compromise between these two conflicting considerations is to select the observation point so that the angle of sight lies between 30° and 45°, i.e the observer should stand between one and one-and-a-half times the tree height away from the tree. Angles greater than 45° must be avoided as the probability of mistaking a side branch for the top of the tree is unacceptably high.

- ☆ Instrument error. All instruments should be checked periodically against some standard or known height and adjust as necessary.
- ☆ Operator and recording error. - Personal error is always likely, e.g.:
  - Incorrect setting of distance or booking of angles and distances, incorrect reading;
  - Forgetting to add on the section of tree below eye level or forgetting to sight to the tree base;
  - Measuring to wrong tip - shaking the tree may help!
  - Difference of opinion amongst observers in nominating the tip of an umbrageous crown.

Measurement of Logs and Fuel wood

Some Terminology

Basal area: the area of a cross section of a stem at breast height

Billet: A piece of round wood about one meter in length usually cut for pulp or firewood

Cull: the portion of a tree stem or log which is unmerchantable

Log: The stem of a tree or a length of stem or branch after felling and trimming (BCFT)

Measurement of Length, Diameter and Sectional Area of Logs

- ☆ The ultimate object of all mensuration activity in forest is to calculate or estimate quantity of wood contained in trees and consequently in crops not only for sale but also for research, predicting future yields, estimating increment to assess return on capital etc.
- ☆ Measurement of felled trees are to determine the quantity of merchantable volume to obtain statistical data that could be applied

to standing trees for the purpose of estimation the yield, to estimate the growing stock and to estimate the increment of woods and forests.

- Volume estimation may be made most accurately when the logs are separated and accessible to the measurer
- A tree, therefore, could be separated into stem wood, which may be further divided into timber and small wood, crown and branch wood.
- Stem wood may be measured after division into sections for obtaining real volume
- The measurement requires length and mid diameter or mid girth except where the tip is measured as frustum of a cone where the diameter or girth at the ends is measured.
- Logs are neither cylinder nor often of any regular geometric shape. Therefore in order to calculate the volume, the shape of a quadratic paraboloid is adopted.
- It is usual to cut the tree into logs due to irregularity in tree tapers.
- The length of the logs depends upon the rate of taper and market requirements.
- As the diameter at the thin end of the log determines the sawn volume that can be taken out of it, the greater the rate of taper, the lesser is the length of the log.
- Another consideration that affects the length of log is the mode of transport
- When the logs are made for calculating volume of felled trees for research work, all logs including the first are of uniformly 3m in length except the top end log which may be up to 4.5 m. But if the end section is more than 1.5m in length, it is left separate rate log.
- Simple tape or a graduated rod can be used to measure the length of a given logs.
- Similarly, diameter tape, caliper and other optical instrument are used to measure the diameter and sectional area of logs.
- Logs ate the round piece with square cut ends. Normally, a log is 6 or 8 ft over in length and suitable for lumber.
- The cross-sectional area or basal is found from the diameter as follows:

$$BasalArea = \frac{\pi d^2}{4}$$

Formulae for Log Volume Calculation

When calculating volumes of logs and trees, we normally assume that the sections are circular, or at least the diameters are such that the area of the section is $\pi D^2/4$.

It is customary in forest mensuration to take the shape of logs and trees as similar to certain solids of revolution, the cylinder, paraboloid, cone, or neiloid.

More generally, different parts from the tree resemble portions of these solids. The crown part, in conifers, tends to the cone form. The stem central part approaches a paraboloid. The base of the tree expands in a form similar to the neiloid, although generally values of *n greater than 3 come closer.*

a. Prismoidal or Newton's formula

$$V = \frac{S_1 + 4S_m + S_2}{6} * l$$

Where,

V = Volume of Logs

S_1 = the sectional area at the thick end

S_m = the sectional area at the middle

S_2 = the sectional area at the thin end

l = the length or height of the solid

- ☆ It is the best and accurate method for volume calculation
- ☆ It gives particularly the volume of frustum of Neiloid (as well as other sections)
- ☆ It is only used to calculate the error in volume calculated by other formula
- ☆ It is difficult to apply particularly when the logs are stacked

b. Smalian's Formula

$$V = \frac{S_1 + S_2}{2} * l$$

Where,

V = Volume of Logs

S_1 = the sectional area at the thick end

S_2 = the sectional area at the thin end

l = the length or height of the solid

- ☆ It gives the volume of frustum of paraboloid (also cylinder)
- ☆ It over estimates the volume
- ☆ It is used when the logs are stacked or lying on the ground

c. Huber's Formula

$$V = S_m * l$$

Where,

V = Volume of Logs

Sm = the sectional area at the middle

l = the length or height of the solid

- ☆ It gives the volume of frustum of paraboloid (also cylinder)
- ☆ It under estimates the volume
- ☆ It is difficult to apply particularly when the logs are stacked.
- ☆ It is easier and more accurate than Smalian's formula.

d. Quarter Girth formula (Hoppus's rule)

$$V = \left(\frac{g}{4}\right)^2 * l$$

Where,

V = Volume of Logs

g = the girth of the log at the middle

l = the length of the log

- ☆ This is the system of measurement used in Great Britain and also in Nepal for sale purpose when round timber is sold by volume
- ☆ This formula gives only 78.5% of the cubic volume of cylinders, thus allowing a loss of 21.5%

Volume of Sawn Timber

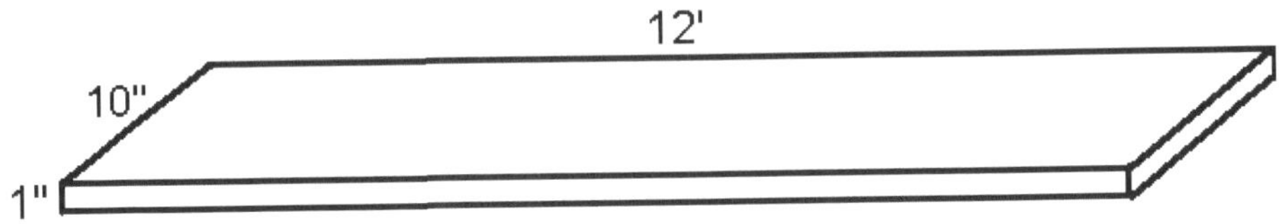

Fig. 28: Sawn timber

- It is the simple product of three dimensions; length, breadth and height
- Volume of sawn timber varies according to its cross-sectional size and length.

Volume of Stacked Timber

Products such as firewood and pulp logs are frequently commercialized according to their volume in piles or stacks.

Fig. 29: Stacked timber

- A stere metre is the volume of a stack of 1×1×1 metres (a cubic metre stacked), and it is used for firewood
- This volume contains air space and wood in variable proportions according to the form of logs
- Piling co-efficient has to be used to get the actual volume
- Piling co-efficient= $\pi/4=0.7854$ if all piece of wood were cylindrical and of the same diameter
- Due to edge effects the wood content can vary slightly with the stack dimensions, and much with the stacking method, so buyers and sellers usually establish specific norms on dimensions and stacking methods.
- Other important factors in the solid content are the irregularity of the logs, the variability of the diameters, and the bark thickness.
- Movement during transport can also introduce important changes.

Dimension and Volume of Chatta (Stacked Fuelwood)

Fig. 30: Chatta

- ☆ Standard size of chatta= 5ft.*5ft. *20ft. = 500 cft including air space
- ☆ One chatta in metric unit= 14.16m^3
- ☆ The following formula should be used in order to calculate the amount of fuelwood that is obtained from the total volume up to 10 cm top-diameter of class III and the remaining portions up to 10 cm top-diameter of class I and II trees which would not be used as timber.

Volume of Chatta

$$\text{Amount of fual woodin terms of chatta} = \frac{(0.8778*\text{Vol.I}+1.4316*\text{Vol.II}+3*\text{Vol.III})}{1000}$$

Where,

Vol I = gross volume of up to 20cm top diameter of class I trees

Vol II = gross volume of upto 20cm top diameter of class II trees

Vol III = gross volume of upto 10cm top diameter of class III trees

All trees except Khair having dbh of 27.94 cm (11 inch) or above should be classified as below

- ☆ **Class I:** Green, dead or dying, standing or uprooted tree having good and solid trunk in which sign of any disease or wound is not visible from outside
- ☆ **Class II:** Green, dead or dying, standing or uprooted tree in which complete volume could not be realized due to hollowness or other sign of defect but at least two straight logs of each 1.83 m (6ft) long or one straight log of 30.5 m (10 ft) long which should have at least 20 cm diameter could be recovered.
- ☆ **Class III:** Remaining trees which do not fall under class I and class II

Solid Volume of Fire Wood

- The stacked volume is not the actual volume of firewood, it is only for the convenience of paying the labour in the forest where there is no arrangement for weightage.
- Solid volume of firewood in a stock depends upon several factors such as care in stacking, form of billets, length of billets and their diameter

Xylometric Method

- Volume of billets calculated with the help of xylometer which consists of a graduated vessel
- Volume of wood calculated by the principle of water displacement
- Water poured in vessel, reading taken, woods are submerged in water and reading taken again.
- Difference between two reading gives the volume of submerged wood

For large quantities of wood

If,

W = Weight of whole stack of wood

w = Weight of submerged pieces

V = Volume of the whole stack of wood

v = Volume of submerged pieces

$$W : w = V : v$$

$$V = \frac{W * v}{w}$$

- This method is cumbersome and seldom used in practice

Specific gravity method

- Specific gravity is a unitless measure of mass.
- If specific gravity of wood is known than volume can be calculated.

$$\text{Specific gravity of a piece of wood} = \frac{\text{Weight of wood}}{\text{Weight of same volume of water}}$$

$$\text{Specific gravity of a piece of wood} = \frac{\text{Density of wood}}{\text{Density of water}}$$

$$Volume = \frac{\text{Weight (gms)}}{\text{Specific gravity}} cc$$

- ☆ As density for pure water is 1 gm per cc, the density of wood in gm per cc is the same as its specific gravity minus the units.
- ☆ Specific gravity typically varies from 0.35 to 0.81 for most commercial tree species

UNIT 3

Measurement of Tree Form

Tree Stem Form

- Form is the rate of taper of a log or stem
- It is the decrease in diameter of a stem of a tree or of a log from base upward.
- The taper varies not only with species, age, site and crop density but also in the different parts of the same tree.
- The basal portion of the tree corresponds to the frustum of a neiloid, the middle portion to the frustum of a paraboloid, and the top portion to the cone
- Trees often are combinations of form.

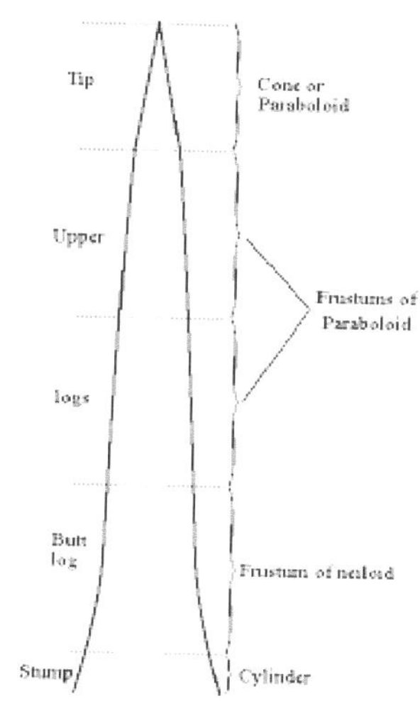

Fig. 31: Tree form

Metzger's Theory or Girder Theory

- Several theories put forward to explain variations in taper from tree to tree and in the same tree as well.
- Metzger's theory assumes that the tree stem should be considered as a cantilever beam of uniform size against the bending force of the wind.

- The wind pressure acts on the crown and is conveyed to the lower parts of the stem in an increasing measure with the increasing length of the bole.
- Thus, the biggest pressure is exerted at the base and there is a danger of trees snapping at the place, to counteract this tendency, the tree reinforces itself towards the base.
- The pressure of wind crown keeps on changing as the tree is growing in open crowded portion.
- Tapering increase if it is an isolated area, an area where largest density, in the area tapering decreases.
- Though tapering is the natural process which can be controlled by human interference. If competition increases, tapering decreases.
- Trees growing in complete isolation or exposed situation have short but rapidly tapering boles while the trees growing in dense crops, which are therefore subjected to lesser wind pressure, have long and nearly cylindrical boles.

Mathematically,

Let,

p= a force applied to a cantilever beam at its free end

l= the distance of a given cross section from the point of applications of this force

d= the diameter of the beam at the point

s= the bending stress in kg/cm2

By the rule of mechanics,

$$S = \frac{p * l}{d^3} * \frac{32}{\pi}$$

As the force p in case of trees consists of components

W= wind pressure per unit area, F= crown area, then p=w*F

Then,

$$s = \frac{w * F * l}{d^3} * \frac{32}{\pi}$$

$$d^3 = \frac{32 * w * F * l}{\pi * s}$$

- ☆ For a given tree w, F, s can be considered as constant, therefore $d^3 = kl$, where k is a constant.
- ☆ Thus, the diameter raised to the third power increases proportionately with lengthening of the lever or with the increasing distance from the central point of application of wind force.
- ☆ According to this logic, the tree stem must have the shape of a cubic paraboloid.

Methods of Studying Form

- ☆ By comparisons of standard form ratios (form factor and form quotient).
- ☆ By classification of form on the basic of form ratios and,
- ☆ By compilation of taper table.

Form Factor and its Type

- ☆ Form factor is the ratio of the volume of a tree or its part to the volume of a cylinder having the same length and cross section as the tree.
- ☆ It is the ratio between the volume of a tree to the product of basal area and height.

$$F = \frac{V}{Sh}$$

Where,

F = form factor

V = tree volume in cubic units

S = basal area at breast - height in area units

h = height of the tree in linear units

Types of Form Factor

a. Artificial Form Factor

- ☆ Known as breast-height form factor
- ☆ Basal area or diameter measured at dbh and the volume refers to the whole tree both above and below the point of measurement.
- ☆ It is not reliable guide to the tree form.

- ☆ Diameter measurement is fixed, but no fixed relation exists to the height of the tree and portion above the breast height.
- ☆ Trees of same form but different heights will have different form factor.
- ☆ Universally used for its handy measurement and standardization of diameter at breast height.
- ☆ A useful application is for quick-and-dirty volume estimates, assuming a constant form factor.

b. Absolute Form Factor

- ☆ It is the ratio between the volume of the tree above the point of diameter or basal area measurement with the cylinder which has the same basal area and whose height is equal to the height of the tree above that point.
- ☆ Basal area is measured at any convenient height and the volume refers to that part of the tree above the point of measurement

c. Normal form factor

- ☆ Basal area is measured at a constant proportion of the total height of the tree, e.g $1/10^{th}$, $1/20^{th}$ etc. of the total height and the volume refers to the whole tree above ground level.

Disadvantages

- ☆ The height of tree to be determined before the point of measurement can be fixed.
- ☆ Point of measurement very inconvenient in case both very tall and short trees.

Absolute form factor and normal factor are no longer used. Unless stated, form factor implies artificial form factor whose basal area calculated at 1.3 m. The natural form factor corresponding to the total volume of a stem is generally between 0.3 and 0.6.

Uses of Form Factor

- ☆ To estimate volume of standing trees
  - Form factor compiled in tabular form to give average form factor of different dimensions by dbh and height classes.
  - Table used to estimate volume by measuring dbh and height.

- Table prepared from measuring large number of trees, so application to individual tree not satisfactory, however used to estimate volume of group trees.
- Uses limited to similar growing conditions

☆ To study laws of growth

- Gives insight to laws of growth, particularly to stem form of trees.

Kinds of Form Factor

Depending upon volume represented, form factor may be of following kinds

☆ Tree form factor

☆ Stem timber form factor

☆ Stem small wood form factor

Form Height

It is the product of form factor and total height of the tree.

$$Fh = \frac{V}{S}$$

Where,

Fh = form height

V = volume of the tree

S = basal area

Volume is calculated from under bark measurements and the basal area is calculated from dbh (ob). Form height is used to determine how far is it reasonable to assume that volume is proportional to the basal area. If form height remains constant with increasing diameter, then it is clear that the assumption is justified.

Form Quotient and its Type

Form Quotient

☆ It is the ratio between the mid-diameter and the dbh.

$$F.Q = \frac{\text{mid - diameter}}{\text{dbh}}$$

☆ Taper depends upon form quotient (A. Schiffel)

Types of form quotient

- ☆ **Normal form quotient**
  - Ratio of mid-diameter or mid-girth of a tree to its diameter or girth at breast height.
- ☆ **Absolute form quotient**
  - Ratio of diameter or girth of a stem at one half its heights above the breast height to the diameter or girth at breast height.

Form quotient is the third independent variable of volume table that can be used to predict the volume of a tree stem.

Form Class

- ☆ Form class is defined as one of the intervals in which the range of form quotients of trees is divided for classification and use.
- ☆ It also implies to the class of trees which fall into such an interval.
- ☆ Trees may be grouped into form classes expressed by form quotient intervals such as 0.50 to 0.55, 0.55 to 0.60 and so on or by mid-points of these intervals such as 0.525, 0.575 and so on.

Form Point Ratio

- ☆ It is defined as the point in the crown as which wind pressure is estimated to be cantered.
- ☆ Form point ratio is defined as the relationship, usually expressed as a percentage, of the height of the form point above ground level to the total height of the tree.
- ☆ Form point ratio bears a consistent relation to the form quotient.
- ☆ If form point ratio is known, the form quotient and form class of a tree can be determined.

Taper Table and Formulae

- ☆ It provides the actual form by diameters at fixed points from the base to the tip of a tree.
- ☆ Volume tables can thus be prepared from taper tables in desired unit.

Use of Taper Table

- ☆ Volume of the average tree for each diameter and height class can be found readily in office without direct measurement. The only measurement that will be needed is the dbh (ob) and the height of standing tree.

- Volume table can be prepared from taper tables in desired units.

The ultimate purpose of all taper tables is to show upper stem diameters, which can then be used to calculate the volume of the sections of a tree and the entire tree. Taper tables can assume several forms.

Types of Taper Table

- **Ordinary taper table or diameter taper table**
  - It gives the taper directly for diameter at breast- height without reference to the tree form.
- **Form class taper table**
  - This tables gives for different form classes the diameters at fixed points on the stem expressed as percentage of dbh (u.b)

General Formulae or Equations for Tree Form

- Taper equation represents the expected diameter as a function of height above ground, total tree height and dbh irrespective of tree species and generalized for form class
- Many different forms of taper equations have been developed as no single one can adequately represent all species in all situations. The use of taper equations allows us to obtain volumes for any desired portion of a tree stem by predicting upper stem diameters.

Hojer's Formulae

- It determines the diameter quotient (i.e the ratio of the diameter of a stem at any given height to its breast-height diameter) for each form class
- It gives the percentage of the length of the tree between breast height and top

$$\frac{d}{dbh} = C \log \frac{c+l}{c}$$

Where,

d = Diameter at any point on the stem

C & c = Constant for each form class

l = Distance from the top of the tree to the point at which d is measured

Behre's Formulae

$$\frac{d}{dbh} = \frac{l}{a+b}$$

Where,

d = Diameter at any point on the stem

a & b = Constant for each form class, a + b = 1

l = Distance from the top of the tree to the point at which d is measured

- ☆ This formula is more consistent.

Bark Measurement

- ☆ The thickness of the bark and its percentage of volume in the tree or log are important parameters in mensuration because most measurements on standing trees have to be made on over bark.
- ☆ Some species have very thick bark. In general, bark thickness varies with: species, age, genotype, rate of growth and position in the tree
- ☆ The bark thickness of the living tree may be measured with little damage to the trees using a Swedish Bark Gauge

$$\text{Bark Percent} = \frac{V_{ob} - V_{ub}}{Vob} * 100$$

Where,

V_{ob} = Volume over bark

V_{ub} = Volume under bark

UNIT 4

Forest Sampling and Inventory

Definition and Scope of Sampling

- Sampling is the process in which enumeration is to be done only in a representative portion of the whole.
- In Sampling, the information is obtained only from a part of the population assuming that it is the representative of the whole. On that basis the conclusion is drawn for the entire population.
- For example, a forest area may be of 1000 ha out of which only 100 ha have been selected for enumeration and estimate of the whole population of 1000 ha is made, it is called sampling.

Sampling unit: The population is divided into suitable units for the purpose of sampling.

Types of sampling units in forest surveys are:

- Compartments,
- Topographical sections,
- Strips of a fixed width,
- Plots of definite shape and size etc.

Sampling frame: The list of sampling units from which the sample units are to be selected is called sampling frame.

Sampling Intensity (SI): The ratio of sample to the whole population which is expressed on a percentage

SI = sample area/Total area x 100

Size of the Sample

The number of sample units in the sample is known as sample size.

Factors affecting the size of sample

- ☆ Nature of population
- ☆ Number of classes
- ☆ Nature of Study
- ☆ Types of sampling
- ☆ Standard of accuracy/precision required, and
- ☆ Other considerations

Scope of Sampling

- ☆ Less time
- ☆ Reduced cost
- ☆ Administrative convenience
- ☆ Better supervision
- ☆ Check result of census method
- ☆ Suitable for infinite/hypothetical population
- ☆ Suitable for destructing sampling

Limitations of Sampling

Sampling is better over complete census only if

- ☆ The sampling units are drawn in a scientific manner.
- ☆ Appropriate sampling technique is used, and
- ☆ The sample size is adequate.

Sampling theory has its own limitations and problems, which are:

- ☆ Proper care should be taken in the planning and execution of the sample survey; otherwise the results obtained might be inaccurate and misleading
- ☆ Sampling theory requires the services of trained and qualified personnel and sophisticated equipment for its planning, execution

and analysis. In the absence of these, the results of the sample survey are not reliable.

- ☆ If the information is required about each and every unit of the universe, there is no way but to resort to complete enumeration.

Types of Sampling

Random and Non-random Sampling:

Probability/random sampling

- ☆ Simple random sampling
- ☆ Stratified random sampling
- ☆ Multistage sampling
- ☆ Multiphase sampling
- ☆ Sampling with varying probabilities

Non random sampling

- ☆ Selective sampling
- ☆ Systematic sampling

Simple Random Sampling

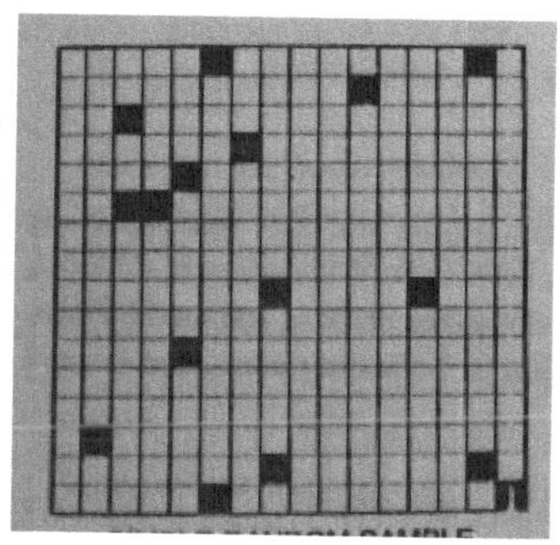

Fig. 32: Simple Random Sampling

- ☆ It is a selection process in which every possible combination of sample units has an equal and independent chance of being selected in the sample.
- ☆ Sampling units are chosen completely at random.
- ☆ For theoretical considerations, SRS is the simplest form of sampling and is the basis for many sampling methods.
- ☆ It is most applicable for the initial survey in an investigation and for studies that involve sampling from a small area where the sample size is relatively small.

Selection of SRS

- ☆ Lottery Method and
- ☆ Random number table method

When to Use

- ☆ If the population is more or less homogenous with respect to the characteristics under study and
- ☆ If the population is not widely spread geographically.
- ☆ 16 samples are selected randomly from a population composed of 256 square plots

Advantages

- ☆ SRS is a scientific method and there is no possibility of personal bias.
- ☆ Estimation method are simple and easy.

Disadvantages

- ☆ If the sample chosen is widely spread, takes more time and cost.
- ☆ A population frame or list is needed.

Stratified Random Sampling

- ☆ It is a method of sampling in which the populations is first divided into sub population of called strata of same or different size in such a way that characteristics within the strata are homogenous but between the strata are heterogeneous.
- ☆ Samples are taken from each stratum by randomly or other method regarding to optimum or proportional allocation methods.

Criteria of Stratification of Forest Area

- ☆ Topographic features, Forest types, Density classes
- ☆ Volume classes, Age classes etc.

1. Proportional Allocation

- ☆ When information regarding the relative variances within strata and cost of operations are not available, the allocation in the different strata may be made in proportion to the number of units in them or the total area of each stratum.
- ☆ Proportion to the area
- ☆ Formula, $n_i = (N_i/N) \times n$

Where, N = total number of sampling units in the population/forests (total population), N_i = the number of sampling units in the ith stratum (stratum size), n_i = the no. of sample units, i = 1, 2, 3, ...k, k = the no. of strata and n = total sample size from all the strata.

- ☆ Larger size strata receive large size sample values.

Example

Sample size(n)= 30

Total population size (N)= 8000

Strata size N^1= 4000, N^2= 2400, N^3=1600

Let P_i represent the proportion of population included in stratum i

Then,

For strata with N1= 4000, we have P1= 4000/8000

Hence,

$n_1 = n^*p_1 = 30(4000/8000) = 15$

Similarly,

$n_2 = n^*p_2 = 30(2400/8000) = 9$

$n_3 = n^*p_3 = 30(1600/8000) = 6$

2. Optimum Allocation

- ☆ Sample plots are allocated to various strata according to standard statistical procedure resulting in smallest standard error possible with a fixed number of observations.
- ☆ Determining numbers of plots to be assigned to each stratum requires first a product of the area and standard deviation of each type
- ☆ Minimize the variance (i.e. maximize the precision) of the estimate
- ☆ Other thing being equal, a larger sample may be taken from a stratum with a larger variance so that the variances of the estimates of strata means get reduced.

$ni = n \times N_iS_i/\sum N_iS_i$

Where, S_i = standard deviation of each stratum

Example

Sample Size (N)= 84

Total Population Size (N)= 10000

Strata Size N_1= 5000, N_2= 2000, N_3= 3000

$\sigma_1 = 15$

$\sigma_{2=} = 18$

$\sigma_3 = 5$

$$n_i = \frac{n * N_i \sigma_i}{N_1\sigma_1 + N_2\sigma_2 + ... + N_k\sigma_k}$$

Where

$i = 1,2,...k$

$$n_1 = \frac{84(5000)(15)}{(5000)(15) + (2000)(18) + (3000)(5)} = \frac{6300000}{126000} = 50$$

Similaarly,

$n_2 = 24$

$n_3 = 10$

When to use – when the sampling units are **heterogeneous** with respect to characteristics under study.

16 samples are selected randomly from a population composed of 256 square plots.

Fig. 33: Stratified Random Sampling

Advantages

- ☆ More representatives than SRS & systematic sampling
- ☆ Greater accuracy than SRS
- ☆ Administrative convenience

Disadvantages

More time & cost.

- ☆ Sampling units for each stratum is necessary or separate frame is needed for each stratum.
- ☆ Need prior & additional information about population & its subpopulation.

Systematic Sampling

- ☆ Systematic sampling is a commonly employed technique if the complete and up to date list of the sampling units is available.
- ☆ In this sampling technique, first unit is chosen randomly and the rest being automatically selected according to some predetermined patterns.

- ☆ In this sampling, the sampling units are spaced at fixed intervals throughout the population.
- ☆ Measure of every tree along a certain compass bearing is an example of systematic sampling.
- ☆ A common sampling unit in forest surveys is a narrow strip at right angles to a base line and running completely across the forest, i.e. systematic sampling by strips.
- ☆ Another possibility is known is systematic line plot sampling where plots of fixed size and shape are taken at equal intervals along equally spaced parallel lines.

When to use- if the complete or up to date lists of the sampling units are available.

16 samples are selected systematically from a population composed of 256 square plots.

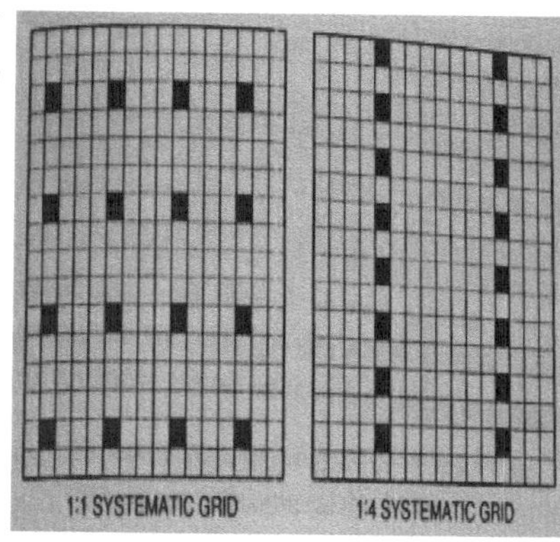

Advantages

- ☆ The distribution of the selected sampling units is controlled by designer and is regular.
- ☆ Field teams understand the layout and can usually identify and locate the units easily.
- ☆ Qualitative and quantitative information over the whole population can be collected at regular intervals during the survey
- ☆ The precision of the parameters estimated is usually high.

Disadvantages

- ☆ No valid estimates of the precision are feasible unless the pattern of variation is known.
- ☆ There is no rational method of choosing the sampling intensity and layout of sample to achieve a desired precision at minimum cost

According to community forestry inventory guidelines,2061

- ☆ **Sampling Method**

Stratified systematic sampling

- **Sampling Intensity**

0.1% (for regeneration and open land), 0.5 % (general forest) in community forest

- **Timber Volume Calculation**

Timber volume = π x d2/4 x ht. x ff x TQ

Where, ff = form factor (0.5), TQ, first quality (by 2/3), second quality (1/2)

Tree Quality (TQ)

- **First quality**- straight and clear bole, 3 or more logs with 6 feet length obtained.
- **Second quality**- straight bole, 2 logs with 6 feet length obtained.
- **Third quality** – crooked and abnormal bole and only fuel wood obtained.

Inventory

Introduction and Scope

- Forest inventory is the procedure of obtaining information on the quantity and quality of the forest resources and many of the characteristics of the land area on which the forest is located.
- Most forest inventories have been, and will continue to be, focused on timber estimation. Thus, Forest inventory is the tabulated, reliable and satisfactory tree information.
- However, the need for information on forest health, water, soils recreation, wildlife and scenic values, and other non-timber values has stimulated the development of integrated or multi resource inventories.
- It is also called enumeration or cruise

An inventory of a forest area can provide information for many different purposes; it may be part of-

- A natural resource survey with the aim of allocating land to different uses, i. e. land planning
- A natural project to assess the potential for forest and wood-based industry development
- A wood-based industry feasibility study
- A resource assessment for forest management planning

- The usual purpose of a timber inventory is to determine the volume (or value) of standing trees in a given area. To attain this objective requires a reliable estimate of the forest area and measurement of all or an unbiased sample of trees within this area.
- The information may be obtained from measurement taken on the ground or on remotely sensed imagery (aerial photographs, satellite imagery etc.).

Types of Inventory

Total enumeration (census): enumeration is carried out over the entire area of the forest unit under consideration. It is expensive and time consuming.

Partial enumeration: enumeration is to be done only in a representative portion of the whole forest.

- The choice of a particular inventory system is governed by relative cost, size and density of timber, area to be covered, precision desired, number of people available for fieldwork, and length of time allowed for the estimate.
- Regardless of the kind of inventory being under taken, a carefully developed plan is needed to execute the inventory efficiently.
- Many forest inventories are carried out using fixed area sample units. These fixed area sample units are called strips or plots, depending on their dimensions.
- Sample plots can be any shape (circular, square, rectangular or triangular), however, square/rectangular and circular plot shapes are most commonly employed.

A strip can be thought of as a rectangular plot whose length is many times its width.

- Population – is an aggregate of objects under study.
- Sample – a finite subset of the population selected from it for the purpose of the study/investigation.
- Sampling unit: The population is divided into suitable units for the purpose of sampling.

Types of sampling units in forest surveys are:

- Compartments, topographical sections, strips of a fixed width, plots of definite shape and size etc.

- Sampling frame: The list of sampling units from which the sample units are to be selected is called sampling frame.
- Sampling Intensity (SI): The ratio of sample to the whole population which is expressed on a percentage

SI = sample area/Total area x 100

Inventory Planning

- Purpose of the inventory
- Background information
- Description of area
- Information required in final report
- Inventory design
- Procedures for interpretation of photo, satellite imagery, or other remote sensing material
- Procedures for fieldwork
- Compilation and calculation procedures
- Final report
- Maintenance

Strip System of Cruising

- With this system, sample areas take the form of continuous strips of uniform width which are established through the forest at equally spaced intervals, such as 100m, 200m, or 300m etc.
- The sample strip itself is usually 20 m to 40 m wide.
- Strips are commonly run straight through the tract in the north-south or east west direction, preferably oriented to cross topography and drainage at right angles.
- By this technique, all soil types and timber conditions from ridge top to valley floor are theoretically interested to provide a representative sample.
- Strip cruises are usually organized to sample a predetermined percentage of the forest area.
- Cruise intensity (I) = W/D x 100,

Where, W =strip width and D = distance between strips

Advantages

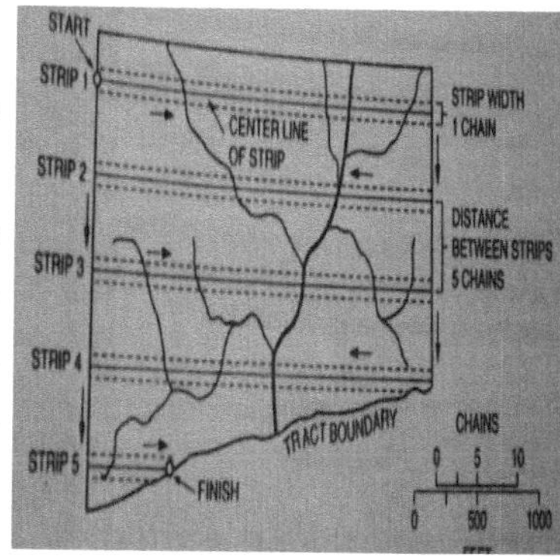

- ☆ Sampling is continuous and less time is wasted in traveling between strips.
- ☆ Strips have fewer bordering trees than plot cruise of the same intensity.
- ☆ With two persons working together, there is less risk to personnel in remote or hazardous regions.

Disadvantages

- ☆ Errors are easily incurred through inaccurate estimation of strip width.
- ☆ It is difficult to make spot checks of the cruise results, because the strip centerline is rarely marked on the ground.
- ☆ Brush and windfalls are more of a hindrance to the strip cruiser than to the plot cruiser.

Line-Plot System of Cruising

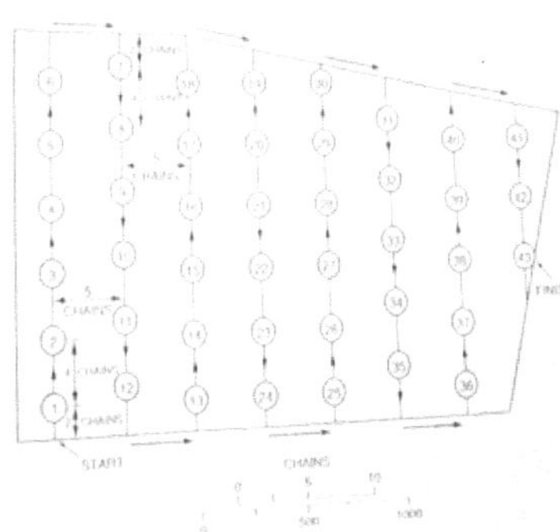

- ☆ Line plot cruising consists of a systematic tally of trees on a series of plots that are arranged in a rectangular or square grid pattern.
- ☆ Compass lines are established at uniform spacing, and plots of equal area are located at predetermined intervals along these lines.
- ☆ Shape of the plots may be circular, rectangular, square or triangles. But circular plot is widely used.
- ☆ Sampling intensity (I)
- ☆ I = (no. of plots x area of plot)/ Total area x 100
- ☆ An intensity of plot sampling is governed by the variability of the stand, allowable inventory costs, and desired standards of precision.

Advantages

- ☆ The system is suitable for one person cruising.
- ☆ Tree tally is separated for each plot, thus permitting quick summaries of data by timber types, stand sizes or area condition classes.

- Cruisers do not have to tally trees while following a compass line.
- Cruiser gets some time, at each plot center to check stem dimensions, borderline trees, and defective trees.

Disadvantages

- It takes more time

Inventory with Point Sampling

- Point sampling is a method of selecting trees to be tallied on the basis of their sizes rather than by their frequency of occurrence.
- It has been found that counting from a random point, the no. of trees whose breast-height cross-section exceeds a certain critical angle when multiplied by a factor gives an unbiased estimate of basal area/ ha.
- This technique is called: angle count cruising, Point Sampling, variable plot cruising, PPS (Probability proportional to size) sampling.
- Sample points are located within a forested tract, and a simple prism or angle gauge that subtends a fixed angle of view is used to sight in each tree diameter at breast height.
- Tree boles close enough to the observation point to completely fill the fixed sighting angle are tallied; stem to small or too far away are ignored. The resulting tree tally may be used to compute basal area, volumes, or numbers of trees per unit area.
- Point sampling can be either horizontal (For basal area estimation) or vertical (For ht. estimation) horizontal sampling has been widely used
- The probability of tallying a given tree depends on its cross-sectional area and the sighting angle used. The smaller the angle, the more stems will be included in the sample.
- PS does not require direct measurements of either plot areas or tree diameters. A predetermined basal area factor (BAF) is established in advance of sampling and resulting tree tallies can be easily converted to basal area per unit area.
- BA conversion factors are dependent or the sighting angle (or critical angle) arbitrarily selected.

Horizontal Point Sampling

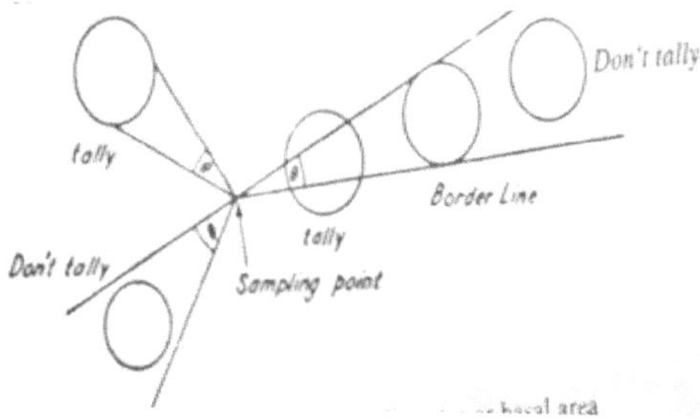

- In horizontal point sampling, a series of sampling points are selected randomly or systematically distributed over the entire area to be inventoried.
- Trees around this point are viewed through any angle –gauge at breast height and all trees forming an angle bigger than the critical angle of instruments are counted.
- Even though all trees are of same basal area, some are counted while others are not because of being far away from sampling points and they do not form an angle bigger than critical angle.
- Inclusion of trees in tally for a given angle depends upon (i) sizes of trees (ii) their distance from the sampling point.
- Basal area per ha = BAF x number of tally trees/ numbers of points
- Number of trees/ha = (no. of trees tallied) x (per-ha conversion factor)/ total numbers of points (per-ha conversion factor= BAF/BA per tree)
- Volume/h = basal area x stand form height

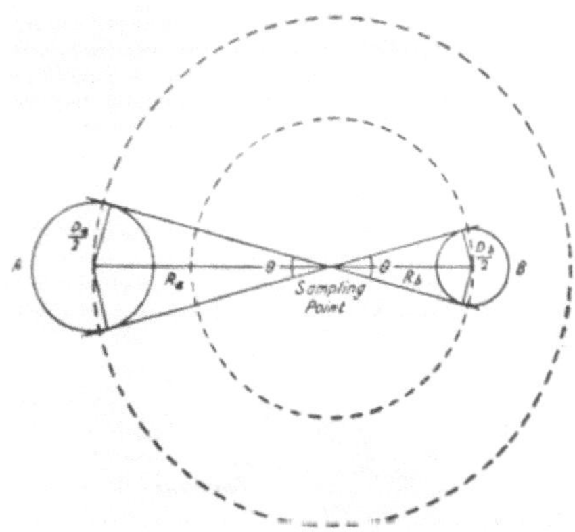

From fig considering the tree section A

$$\text{Sin } \theta/2 = (D_R/2)/R_R = D_a/2R_R$$

Similarly, considering the tree section B

$$\text{Sin } \theta/2 = (D_b/2)/R_b = D_b/2R_b$$

$$2 \text{ Sin } \theta/2 = (D_b/2)/R_b = D_b/2R_b$$

$$K = D/R = 2 \text{ Sin } \theta/2$$

Instruments Used in Horizontal Point Sampling

- Angle gauge
- Wedge prism
- Spiegel Relaskop
- Tele Relaskop

Choice of Instruments

- When steep slopes are regularly encountered, the Spiegel Relaskop is preferred
- For relatively flat topography, either the wedge prism or the stick gauge may be used.
- The prism is primarily desirable for persons who wear eyeglasses.
- The simple stick gauge is preferred in dense stand.

Intensity of Point Sampling

- There is no fixed plot size when using point sampling.
- Each tree has its own imaginary plot radius (depending on the BAF used), and the exact plot size cannot be easily determined, even after the tally has been made.
- However, approximations can be made on the basis of the average stem diameter encountered at a given point.
- From the statistical point view, however, the selection of trees according to size rather than frequency may more than offset this reduction of sample size and with an additional saving in time.
- Conversely, it must be remembered that smaller samples of any kind require larger expansion factors.
- Thus, when point sampling is adopted, the so-called borderline trees must always be closely checked.
- The only accurate method of determining how many sample points should be measured is to determine the standard deviation (or coefficient of variation) of BA or volume per ha from a preliminary field sample.

If the statistical approach is not feasible, the following rules of thumb will often provide acceptable results:

- If the BAF is selected according to tree size so that an average of 5 or 12 trees are counted as each point, use the same number of points.
- With a BAF 10 angle gauge and timber that averages 12 or 15 in. in diameter, use the same number of points.
- For reliable estimates, never use fewer than 30 points in natural timber stands or less than 20 points in even-aged plantations.

Advantages

- ☆ It is not necessary to establish a fixed plot boundary; thus, greater cruising speed is possible.
- ☆ Large high value trees are sampled in greater proportions than smaller stems.
- ☆ BA and volume per ha may be derived without direct measurement of stems.
- ☆ When volume per ha conversions are developed in advance of fieldwork, efficient volume determinations can be made in a minimum of time. Thus, the method is particularly suited to quick, reconnaissance type cruises.

Disadvantages

- ☆ Heavy underbrush reduces sighting visibility and cruising efficiency.
- ☆ Because of the relatively small sized of sampling units, carelessness and errors in the tally (when expanded to tract totals) are likely to more serious than in plot cruising.
- ☆ Slope compensation causes difficulties that may result in large errors unless special care is exercised.
- ☆ Unless taken into account, problems can arise in edge-effect bias when sampling very small tracts or long, narrow tracts.

Vertical Point Sampling

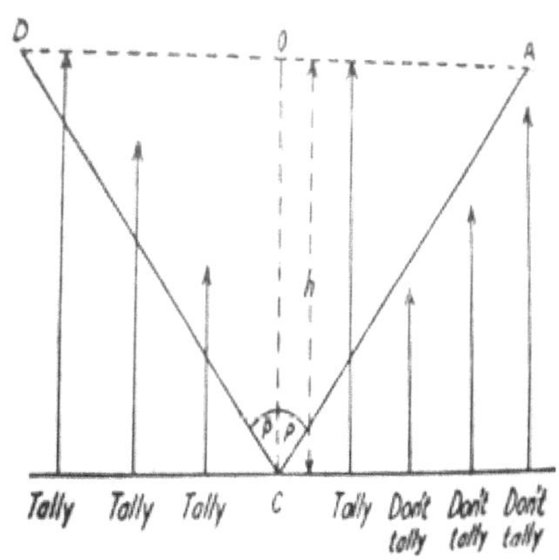

- ☆ A method for deriving the mean stand height
- ☆ Within a full 360^0 sweep around the sample point all trees appearing taller than a critical angle are counted.
- ☆ Instrument used – Conimeter
- ☆ This instrument subtends a critical angle of 45 degree.
- ☆ By using Conimeter, if the number of trees per ha is N, then number of trees counted (n) in the area can be counted from the Conimeter.
- ☆ The average height of the trees (h) in meter (when the critical angle is 45 degree) is

$$h = \frac{100}{\tan p}\sqrt{\frac{n}{\pi N}}$$

$$= 56.4\sqrt{\frac{n}{N}}$$

For crop height, the height at which the instrument is kept will require to be added.

Inventory of Important NTFP found in Tarai, Mid-hills and High Mountain, in each Locality

1. Introduction

Non –Timber Forest Products (NTFP) can be defined as all goods derived from forests of both plant and animal origin other than timber and firewood. Non-wood forest products consist of goods of biological origin other than wood, derived from forests, other wooded land and trees outside forests (FAO 1999). Mallet (1999) defined NTFP as all products, with the exception of timber, that can be harvested from a forest ecosystem.

Resource assessment is an evaluation of some aspect of the resource based on information gathered from a variety of sources. It can include socio-economic issues, market issues, or the quantity and quality of the resource (FAO 2001)

Inventory is an itemized list of current assets (finished goods, components or raw material on hand) (Lund 1997). This may or may not include the quantity or other characteristics of the assets. Forest inventory is a sample-based survey of the forest resource (Burkhart & Gregoire 1994), the intention being to quantify the abundance of biological resources in the forest. The species lists are a subset of an inventory of biodiversity. Thus, a listing of all plants and animals in a forest is not a NTFP assessment but a biodiversity inventory. NTFPs are a sub-set of diversity that has utility to humans. Therefore, the first step in the assessment of NTFPs has to be the identification of a plant or animal whole or part that is collected for some particular use.

It is agreed that the objective of NTFP assessments is to collect information about the distribution, frequency and seasonality of NTFP so that plans can be made for the sustainable management and utilization of the NTFPs for the improvement of human welfare. It is important to determine, prior to the assessment, for whom is the information being collected-who are the beneficiaries/clients. It is also important to determine that the policies and regulations governing the ownership, management and utilization of NTFPs are conducive to supporting farmers and the rural community and hence the local economy. There is no single technique for assessing NTFPs, because of the variety of products. The appropriate methods therefore depend on the objectives of the inventory. The following considerations are important.

The intended use of the inventory: The purpose may be to identify a management plan for an area, or it may be the conservation of an endangered species, or the development of specific products for marketing and processing.

Spatial scale: Inventories may be done at a local, district, national or regional scale, depending on their objectives. Local-level inventories are usually management or market oriented, while regional inventories are usually for planning or policy formulation.

Forest inventories are usual oriented towards specific products-most commonly timber, but increasingly also for some non-timber forest products such as bamboo, medicinal herbs and fruits. The incorporation of many different products in one inventory is complicated by differences in species phenology, scale and spatial distribution. Inventories of non-timber forest products can be purely to determine the quantity of the product available or to determine the quantity used.

The latter involves market studies and the flow of products, which can be age or size dependent. In many cases the biophysical resource inventory would focus on individual products (e.g., bark, fruits), while a socioeconomic flow inventory would concentrate on types of product usage (e.g., food, medicines, gums). The need for an established methodology for NTFP inventory was recognized by a Working Group at an FAO meeting on NTFPs (FAO 1996) which attempted to define a standard for assessing NTFPs. However, the Group concluded that it is 'virtually impossible and therefore perhaps futile to search for a generalized technique for NTFP resource assessment' because of major differences in:

- Intended use of the survey results,
- The groups of NTFPs to be included,
- Spatial scale, and
- Temporal scale.

To this list the resources available for the inventory in terms of technologies, skill levels and finances should be added. Despite these complexities inventory is done for the sustainable harvesting of NTFPs, monitoring the state of resources, considering promotion of resource-based industries and informing conservation of endangered species. Though the inventory methods of NTFP are different, however they follow a basic strategy for managing on a sustained yield basis (Wong 2000).

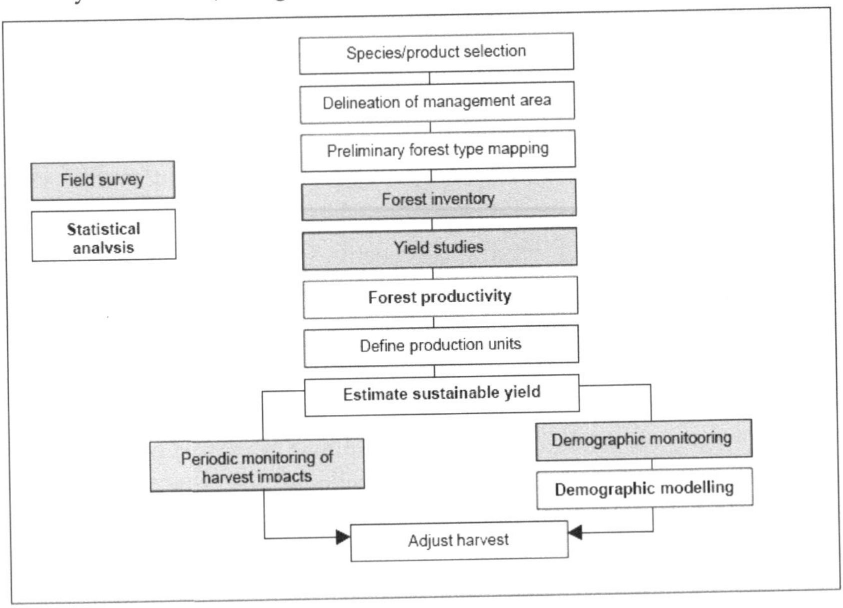

Fig. : Flow chart of a basic strategy for managing NTFPs on a sustained-yield basis (After Peters 1994)

2. Inventory methods for NTFPs

For biologically sustainable exploitation of the products to be determined, there must be a minimum set of good information available on the resource species condition, abundance, distribution and reproductive biology. Resource assessment can be done either qualitatively or quantitatively. In qualitative assessment, based on the ocular judgment/estimation condition of the resource is assessed as abundant, moderate or poor.

Campbell (1989) describes quantitative ecological inventory as 'enumeration of individuals and species of trees in a small patch of forest, the measurement of several important parameters of those individuals and the analysis of the abundance and distribution of those individuals as functions of their physical and biotic environments'. This is subtly different from the foresters definition of 'a sample-based survey of the forest resource' (Burkhart & Gregoire 1994) because of the emphasis on ecology in the former and on sample based observations in the latter. For the purposes of this study quantitative inventory is defined as a biometrically rigorous enumeration of the abundance and distribution of resource populations.

Since NTFPs can be plants or animals it is necessary to review methodologies for each. Although, at first glance these look very different they have the same underlying structure. This structure is envisaged as a hierarchy of design features. At the highest/largest level is the sampling design itself, whether the plots are to be located using random or systematic, stratified or uniform layouts etc. The next level down is the plot scale at which decisions about plot dimensions have to be made. Conventionally, the term plot is taken as meaning a measured area on the ground. However, observations at a point are area-less and therefore one-dimensional, also timed observations (e.g. for bird calls) are defined in terms of time rather than area. 'Plot' should therefore be understood to be any observational unit for enumerating the target species at the local scale (i.e. over metres and hours). Within each plot the enumeration that is undertaken is dependent on the target and product being investigated.

The status of NTFPs in an inventory largely determines how closely methodology is adapted to the peculiarities of the target species and products and the level of resources available. Single resource studies should be best placed to undertake methodological development of specific protocols for assessment of enumeration, this seldom happens. Inclusion of NTFPs into multipurpose resource inventories places very severe constraints on NTFP methodology though there are opportunities which often are not utilized. Methodological studies of optimal inventory design for a range of plant NTFPs are rare. In contrast, methodology for animal inventory is already well developed and requires application rather than further methodological work (Bibby & Moss 1998: Cited in Wong 2000). For the purposes of this study, three contextual settings for NTFP inventory are distinguished. These are single resource, single purpose multiple resource and multi-purpose resource inventories. The distinction between these is the focus of the inventory and the extent to which protocols can be tailored to specifics of the NTFPs in question. Single resource studies are focused exclusively on a particular product and provide the best opportunity for development of tailored protocols and should set the standards for NTFP studies. Single-purpose multi-resource inventories are generally designed to provide management information on NTFPs for a defined area and have as their focus the potential of the land in terms of the species and quantities of NTFPs present and they use land-centric methodology, and differ from multipurpose inventory in that they are undertaken by one institution within the context of a management structure. Multi-purpose resource inventories (MRI) are often multi-institutional and NTFPs are often a small component of the inventory and consequently the development of protocols is heavily constrained by the need to compromise with the needs of other elements of the inventory.

Single Resource Inventory

The objective of a single resource inventory is the quantification of the abundance and distribution of a single product. Such an inventory is unfettered by considerations of other products or purposes and one would expect methodology to be closely tailored to the characteristics of the species from which the product is derived. However, there are very few single product studies that have quantification of the insitu resource as an objective. This could be because a NTFP has to be either very valuable, or subject to legislation for it to justify a species-specific inventory. Therefore most species specific studies have been done for species that are traditionally important for export such as rattan. Although there are few true inventories for single species there are a number of studies that use inventory methodology to address resource related questions and it is these which are examined here.

Six reasons for undertaking a single resource inventory which are (Wong 2000):

1. To provide knowledge of the effects of exploitation of a species for which no other work has been done
2. To assess the potential of particular products for which increased commercialization is sought at either the national or local level
3. To assess the potential of a specific area for exploitation of a known commercial product
4. To investigate the spatial distribution of an exploited product
5. To provide supporting data for the determination of quotas, this is required for several products under national legislation or international treaty
6. For academic inquiry e.g., historical understanding of role of wild yams in historical human diet (Hladick & Dounias 1993, Cited in Wong 2000).

Single Purpose Multi-resource Inventory

The single purpose inventories are intended to provide quantitative information for the purpose of management planning. As such they are area based and attempt to record the presence and abundance of a range of species that occur on a particular portion of land. In selection timber management the forest is divided into compartments (~ 100 ha – 1 km^2). Immediately before logging the compartment is subjected to a stock survey which involves the location, identification, numbering and measurement of every timber tree in the compartment. Yield formulae are used to determine the sustainable yield of timber from the compartment based on the stock survey data. Trees are selected from the stock survey map to make up the allocated yield according to rules designed to protect the environment and the future potential of the forest. Stock survey protocols are generally 100% surveys with the area covered by consecutive enumeration strips. The integration of NTFPs into conventional forest management has resulted in attempts to incorporate NTFPs in timber stock surveys. For example, in Belize, Smith (1995) experimented with a 10% random sample of 1 ha stock survey blocks for a range of products including thatch, edible nuts, palm heart, chicle, incense resin, all spice and decorative epiphytes (Orchids and Bromeliads). Although it is difficult, in the absence of any data to make a judgment of this methodology, it would seem to be a sound and pragmatic means of discovering the distribution, abundance and NTFP management potential of the area to be logged. The only disadvantage is that non-tree NTFPs may be adversely affected by logging and consequently not be available after logging and the stock surveys would need to be complemented by species specific studies to determine optimal sustainable harvesting strategies for each species. This information could then be used to develop management plans for all resources available in each compartment. Detailed information on a range of resources including timber, other plant products and bush meat has been acquired by the Adwenase and Namtee communities using techniques derived from those used for timber stock survey (Gronow & Safo 1996, Cited in Wong 2000). A trial of a method designed to permit quantification of key plant resources to support National Park management planning was undertaken in the Bwindi Impenetrable National Park in Uganda (Cunningham 1996a). The design is based on a few small sample plots and uses basic forestry techniques to enumerate the number and size distributions of trees and bamboo. This methodology is intended to provide information to support management decisions but the small number of replicates leads to imprecision and incomplete coverage leads to bias and inaccuracy suggesting that the information is not adequate for detailed planning and is best used as a guide for strategic planning.

Multi-purpose Resource Inventory

Many resource assessments for NTFPs take place as a component of a multi-purpose resource inventory (MRI). MRIs have been defined as 'data collection efforts designed to meet all or part of the information requirements for two or more products, functions (such as timber management and watershed protection) or sectors (such as forestry and agriculture)' (Lund 1998b). NTFPs originate from forests which generally fall under the jurisdiction of a state forest authority which often have statutory responsibility for maintaining up-to-date forest resource data. Once the forest authority has accepted a responsibility for specific NTFPs they are generally included in routine resource inventory and monitoring. For example, the US Forest Service now includes special forest products as part of forest resource plans in accordance with the National Forest Management Act, the National Environmental Policy Act etc. (Molina *et al.*1993: Cited in Wong 2000). In addition, Lund (1998a) lists further national scale MRI with NTFPs for the Russian Federation. In northern and eastern Europe NTFPs in the form of berries and mushrooms have been recognized and included in recurrent inventory (statutory national inventories usually based around compartments undertaken every 10 years) for almost a hundred years. In Russia industrial scale interest in NTFPs led to the development of a journal (Plant Resources Journal established in 1967) and technical committee (Committee of Wild Berry Research established in 1975) dedicated to plant resources (Rutkauskas 1998: Cited Wong 2000). Rabindranath and Premnath (1997) suggested basically two methods of NTFP assessment based on source of information obtained.

Tapping Local Knowledge

Local people are knowledgeable and can provide information of the species identification, species location, different use and changes in the vegetation pattern. The medium for obtaining such information includes:

Interview

Rural people are the source of information regarding the use, distribution and harvesting of NTFPs, their personal knowledge can be explored through interview to give an overview of the resources. Alexiades (1996) defines four types of interview.

- ☆ **Informal interview**: A completely informal interaction between two or more people. The researcher might take notes during or after the 'interview'.
- ☆ **Unstructured interview**: A discussion where one person is gathering information from another on a particular topic but no specific or guiding questions are used after the initial enquiry (e.g. 'tell me about the plants in the forest'). The informant(s) is aware that it is an interview but the choice of topics to discuss and the direction of the conversation are mostly controlled by the informant.

- **Semi-structured interview**: The interviewer has a list of open-ended questions or topics to discuss (usually 6-10 key questions). The amount of information provided on a topic is largely controlled by the informant and the interviewer is free to follow leads. This type of interview is generally used once specific research topics have been identified.
- **Structured interview**: The interviewer has a fixed length questionnaire. Questions are not open ended but have a limited range of possible responses. Understanding what would be a locally meaningful question is a pre-requisite to designing a questionnaire form. This is because inappropriate closed questions will invariably introduce interviewer bias. Formal surveys are generally used to obtain quantitative information regarding a problem. Nichols (1991) states that 'In order to design a good structured interview survey, you need a full knowledge of the problem you are studying. This in itself limits their use. In a new area one needs methods that are more suitable for exploratory.

Interest Group Discussion

People who share particular set of interests make up an interest group, also called focus group. An interest group can be determined by differences in age, gender, ethnic group, wealth, belief etc. Interest group meetings are conducted to get the people with similar sets of interests to discuss the matter and decide the use and distribution of the resources.

Other Participatory Rural Appraisals (PRA)

PRA is qualitative method of resource assessment which is based on people's view on status and trend of a species of interest. Different PRA tools give different information of a species. Timeline trend, resource mapping, seasonal calendar and matrix ranking are important to assess the species.

Direct Measurement

The vegetation may be heterogeneous or homogenous. Depending upon the condition two methods are applied.

Census Survey or Total Enumeration

Census survey means that enumeration of desired NTFPs over the entire area of the forest unit under consideration. Since it is expensive and time consuming; it is applied when the area of study is small (<5ha) and the NTFP is commercially valuable. Census surveys are often carried out to check on the statistical results of partial enumeration.

Partial Enumeration

The area is partially enumerated and the enumerated samples represent the whole population under study. Partial enumeration can be plot method or plot less method.

Nestling Plot Method

It is most common method employed for sampling various types of the vegetation (Rabindranath 1997). A plot may be rectangle, square, circle or any other shape. Plots may be again taken randomly or systematically and stratified or non-stratified. In Nepal generally rectangular and circular plots are used in community forestry inventory. Nested plot designed are commonly practiced for tree, shrub and herb. The size and number of plots depend on the objective of the study. Ideally 10 % to 15% of the total area should be sampled (Rabindranath 1997). If the area is too large this may not be possible, in this case a minimum of 1 ha should be studied. The sample plot should of course be located at randomly chosen spot. All vegetation falling inside the plot and its replication are studied. The Fig. below shows a rectangular nested plot design commonly adopted in inventory of community forest (25*20m^2 for mature trees, 10*10m^2 for pole, 5*5m^2 for shrub and 1*1m^2 for herbs).

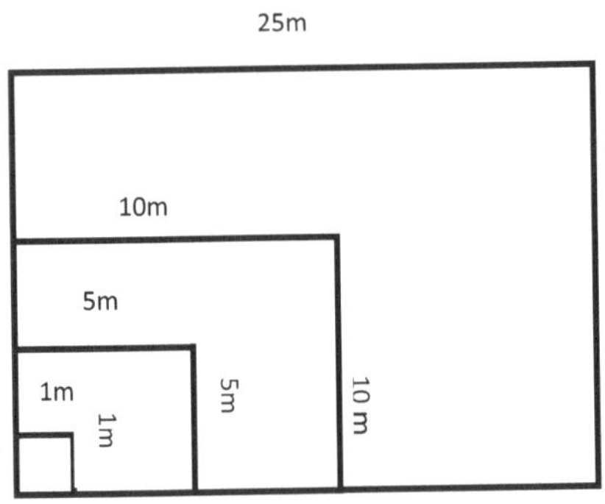

Plotless Method

The plotless methods are best suited for large vegetation area having the heterogeneous plant population (forest area and uneven terrain), where laying plot becomes very difficult. There are several plotless methods however Point Centered Quadrant method is widely used. This is simple and requires less time and more suited for randomly distributed vegetation. In this process a number of points are selected randomly depending on the size of the area, from each point four lines are drawn perpendicularly in all four directions

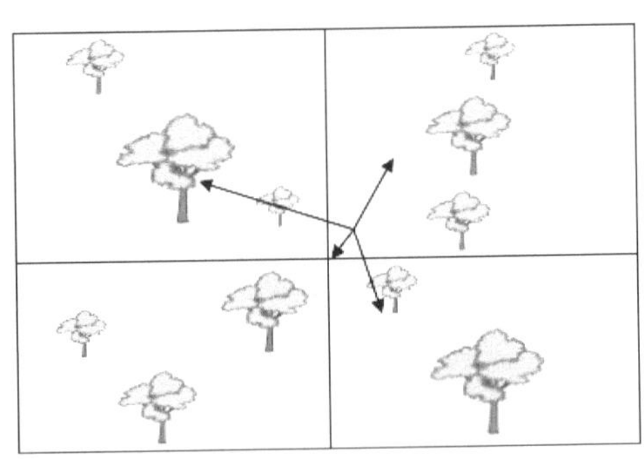

which results four quarters around the point. The nearest tree, shrub or herb is taken as the sample. The following Fig. illustrates the method.

Productivity Assessment

Productivity is the amount of height, girth, volume, biomass that is accumulated by a plant species in certain period of time; may be seasonal, annual or over its whole life span. Different inventory methods are used for different plant parts to determine the productive capacity. These inventory methods may involve the destruction of the whole plants or its parts or without destruction. Resource assessment and productivity of the plants can be done according to the convenience but few important things should be considered.

- ☆ One time resource inventory is not enough
- ☆ It should be done when the leaves are green for easy identification of plant
- ☆ When the products are ready for harvest for productivity estimation of NTFPs
- ☆ More than one period of data collection may be required within a year because different products are ready for harvest at different times of year.
- ☆ Productivity should be measured over several years, rather than just one year.

The inventory methods for fruits, root, leaf, bark taking one example of species is dealt here.

Inventory process of Amla (*Emblica officinalis*) fruits

Amla is a medium sized deciduous tree, but sometimes grows quietly and attains up to 19m height. In Nepal it is found in terai and subtropical valleys from east to west between 100-1600m. Fruits are commercially used. Brief inventory method of Amla is as described (Kunwar 2006).

- ☆ With the participation of local people and reconnaissance survey the distribution of Amla in the forest is determined
- ☆ Forest should be divided into blocks and sub-blocks on the basis of area, geo-structure, natural condition, management objectives, population of species
- ☆ Sampling intensity of 1.5% of site should be taken.
- ☆ Stratified systematic sampling method is adopted while surveying.

- ☆ Necessary inventory plots should be kept equally in every block where Amla is found
- ☆ 25*20 m rectangular inventory plot should be made at every 100 m distance at both right and left sides of the transect line
- ☆ Height and diameter of the trees are measured. Availability, density, etc are calculated.
- ☆ After calculating the density of Amla trees, girth classes (less than 10 cm, 11 to 30 cm, 31 to 65 cm, 66 to 100 cm and above 100 cm) of trees are computed.
- ☆ 5/5 trees per block from each girth class are selected (3/3 trees if there are less than 5).
- ☆ Select 5/5 (or 3/3) tree branches of uniform sizes from each of the upper, middle and lower canopy cover of tree.
- ☆ Pluck or collect the fruits from each branch and weigh the fruits. The mean weight of the fruits from each branch, each tree, each girth class, and each block is determined. This calculation gives the amount of Amla fruits per hectare and that of entire forest.

Inventory of Kutki (*Neopicrorhiza scrophulariflora*) root

Kutki is a perennial herb distributed in the North-Western Himalayan region from Kashmir to Kumaun and Garhwal regions in India, Pakistan, Nepal, Myanmar, Southeast Tibet, North Burma and China (Pollunin and Stainton, 1984: Cited in Giri 2006). It grows naturally in alpine and sub-alpine regions on rock scars as well as in organic soils. It prefers moist habitat formed by winter snow or depressions mostly on the exposed north facing slopes.

In Nepal, Kutki is distributed abundantly also in Sub-alpine and alpine Himalayan regions between 3000-4800m-elevation range with associated of *Rhododendron anthopogan*, Raktelo. It found in the eastern and central regions of Nepal, but most abundantly in the western region where it grows on the rock's crevices on the north facing slopes, cliffs and the turf of glacial flats. Kutki prefers a moist habitat and sandy loam soil (Shrestha and Shrestha, 2004).

It is mainly found in the following districts of Nepal-Taplejung, Sankhuwasabha, Sindhuplancok. Nuwakot, Rasuwa, Ramechhap, Bahlung, Myagdi, Lamjung, Gorkha, Jumla, Mugu, Jajarkiot, Rukum, Bajhang, Bajura, Darchula, Humla, Mustang, Manang, (HMGN, 2055,2058: Cited in Giri 2006). Roots and rhizomes are used indigenously and scientifically. Inventory of Kuti can be done in the following way (Kunwar 2006).

- ☆ With the participation of local people and reconnaissance survey the distribution of Kutki in the forest is determined
- ☆ Forest should be divided into blocks and sub-blocks on the basis of area, geo-structure, natural condition, management objectives, population of species
- ☆ Sampling intensity of 0.5-1% of site should be taken.
- ☆ Stratified systematic sampling method is adopted while surveying.
- ☆ Draw a baseline to the survey block and draw parallel horizontal and vertical lines at a distance of 50m.
- ☆ Lay 2*2 m quadrates at a distance of 10m in each horizontal line.
- ☆ Study the availability, density and fresh weight of Kutki of all quadrant inventory quadrant plots.
- ☆ At the time of maturation leaves become yellow and plants wilt. Roots should be dug out from mature plants.
- ☆ The collection is fresh weighed, sun dried for one week and weighed again to its dry weight. From the difference between fresh weight and dry weight, the conversion factor can be found out. The conversion factor of kutki roots is 0.61.
- ☆ The productivity of kutki of definite location can be calculated by multiplying the conversion factor per hectare with area of forest surveyed.

Inventory of Lokta (*Daphne bholua*) bark

Daphne bholua (Kalo lokta) is an evergreen, erect standing shrub found at an altitude range of 1800m to 3600m. The inner bark of the plant is used for the local manufacture of extremely durable hand-made paper. Lokta paper making in Nepal dates back to at least the 12th century (Trier, 1972: Cited in NSCFP 2001) and continue as an important and rapidly growing cottage industry. The increasing demand of lokta makes its resource inventory an essential for its proper management. The inventory of lokta can be done in a participatory method (NSCFP 2001).

- ☆ **Delineation and estimation of the potential area for lokta**: From participatory mapping, focus group discussion and key informants survey the distribution of lokta in the forest, availability, harvesting suitability is assessed. The potential lokta areas are delineated, block division and estimation of the area of delineated block is conducted.

- Field visit is done to verify the accuracy of participatory mapping, area estimation and preliminary assessment of condition of lokta. During the field visits, parallel transects are constructed covering the whole lokta area. Transects are walked, lokta distribution observed, verify and correct the boundary of potential area. By visual observation, estimate the area of the blocks, cross check it with the obtained values of areas estimated before. Final delineation of the different blocks of lokta and final estimation of their area is done.
- **Defining the resource condition of lokta**: This step involves common understanding of the different qualitative and quantitative resource condition of lokta. By people's participation, stocking condition of lokta can be defined and classified in four resource conditions namely good, fair, poor or nil. The users are asked to keep a visual picture of each of the condition in mind.
- **Estimation of the availability of lokta plants and lokta bark in sample plots**: This step involves the quantification of the amount of harvestable bark, according to different resource condition. Walk along parallel transects lines spaced 100m apart. Stopping at each 20 m interval, resource condition in plots of 5m x 5m is defined as classified previously. In this way we achieve a sampling intensity of about 1%. After observing several plots of each resource condition, a sample of them is selected and measured to assess the amount of harvestable bark and regeneration of lokta. Either randomly or systematically, 10m x 10m plots are taken, regeneration/plant condition measured and classified. Within these plots the plants above the agreed harvestable height are harvested according to the traditional practice; the external bark is separated from the collected product which is weighted (Kg). This helps to quantify the availability of harvestable lokta (number of stems, height of the stems and the weight of the harvestable amount of bark) in each starta or, and so representing each resource condition. This should be done at least 6 times to get reliable estimates.
- **Data analysis**: The total condition of the resource is estimated by multiplying the block area with the percentage spots with resource condition.

Inventory of Machhino (*Gaultheria fragrantissima*) leaf

Machhino is a robust shrub of up to 2m height with ovate to lanceolate evergreen leaves. It grows in open forests and shrub land between 1500 and 2700 m above sea level. It prefers moist and sunny places. The oil extracted from the leaves of Machhino is used as massage oil, as a flavoring agent by the confectionery industry and in the manufacture of soft drinks. The inventory method of machhino as outlined by NSCFP 2001 is as below.

- ☆ **Delineation and estimation of the potential area for machhino**: From participatory mapping, focus group discussion and key informants survey the distribution of machhino in the forest, availability, harvesting suitability is assessed. The potential machhino areas are delineated, block division and estimation of the area of delineated block is conducted.
- ☆ Field visit is done to verify the accuracy of participatory mapping, area estimation and preliminary assessment of condition of machhino. During the field visits, parallel transects are constructed covering the whole machhino area. Transects are walked, machhino distribution observed, verify and correct the boundary of potential area. By visual observation, estimate the area of the blocks, cross check it with the obtained values of areas estimated before. Final delineation of the different blocks of machhino and final estimation of their area is done.
- ☆ **Inventory of the harvestable amount of machhino**: This step involves the quantification of the harvestable raw materials in a number of sample plots. Walk along parallel transects lines spaced 100m apart. Stopping at each 50 m interval, resource condition in plots of 5.64 m radius is defined as good, medium or poor. In this way we achieve a sampling intensity of about 2%. The inventory of Machhino may be also integrated in a general forest inventory, using the same sample plots for the inventory of other products (fodder, fuelwood, litter, ground grass etc.). Once the quality of the shrub has been defined, its average crown diameter has to be estimated by measuring its maximum and minimum diameters and recording the data. In order to simplify this measurement, the borer of the crown can be visualized with a colored rope. The maximum and minimum diameters can then be easily measured with a measuring tape.
- ☆ **Data analysis**: The amount of harvestable machhino is estimated in Kg of fresh material. For this purpose a correlation factor has been defined, that relates the average diameter of the crown and the quality of the shrub to the fresh weight of the harvested product. The total amount of harvestable machhino in the plots is the sum of the fresh weight of all crown diameter classes (excluding those below the agreed minimum harvestable size).

Gaps in the inventory methods of NTFPs

NTFP inventory is complex, there are no particular systems. More over methods in one locality or time may not be appropriate in other locality. Major gaps in NTFP inventory can be outlined as below (Wong 2000).

Problems with adopting traditional forest inventory techniques for NTFPs

The problem of a lack of methods and protocols for NTFP inventory could be easily solved if it were possible to use traditional forestry methods as these are well described and understood.

Unfortunately, this is not a practical solution as there are certain characteristics of NTFPs that are not easily accommodated in traditional forest inventory. The main pitfalls are:

- ☆ **Rarity** - many NTFPs are rare which means that only a few plots of a conventional Systematic or random inventory will contain the species of interest.
- ☆ **Imperfect detectability** – people dealing with trees have rarely come across the problem of searching for an elusive or moving target because trees are generally easy to find. Unfortunately, many NTFPs are not that easy to find and these require that detectability is considered. This is an area where NTFP studies can borrow wildlife techniques.
- ☆ **Seasonality** - Many NTFPs are seasonal but in comparison, timber quantities are constant, consequently forest inventory methods do not cope well with seasonality.
- ☆ **Motility** - Animals run away, fruit falls off a tree and rolls down a hill but trees are static. This is an area where NTFP studies may be able to borrow wildlife techniques.
- ☆ **Quantification of yield for non-destructive harvesting** - most of the methods for determining timber yield from a forest are concerned with the harvesting of entire individuals. For NTFPs one is often only harvesting a small part of the individual. The review suggests that there is little theoretical background for determining harvesting levels of parts of a plant.
- ☆ **Incorporation of local knowledge** – most NTFP studies value local knowledge but there is at present no formal means of using this knowledge to optimize inventory designs. The fact that many studies have struggled with the problem of linking local and scientific knowledge suggests that this would be fruitful area for research.

Lack of Properly Researched NTFP-Specific Sampling Designs

There are relatively few research studies that have looked at optimizing sampling designs specifically for NTFPs. Most of these studies have been done on rattan and, unfortunately, several of these had their own failings. These were mostly to do with plot independence (the trial plot sizes and shapes were contiguous and from only one study site).

Little Guidance Available on Development of Appropriate NTFP Measurement (Mensuration) Techniques

Even for the most commonly studied products there is little advice or standardization of methods.

No Application to NTFPs of Sampling Designs Tailored to Monitoring Needs

There has been no application to NTFP monitoring of any of the statistical thinking on monitoring. Monitoring from a statistical point of view is complicated as there is a problem with controlling the power of the design to discriminate change. Precise observations are required at each time interval to actually detect changes; if the data is imprecise the changes may reside within the errors of the estimates and mask any real change. There has been quite a lot of work done on how to address these problems but to date they have not been applied to NTFPs. Precise monitoring is often very expensive therefore many NTFP monitoring studies use indicators to track changes in the resource. However, there is little known about the linkages between indicators and the resource base. Where this has been studied it has often been shown that a change in off-take or harvest level (a commonly used indicator) is not directly linked to resource condition. What is required are some relatively easy means of determining and tracking change in the relationship between the chosen indicator and resource condition.

Difficulties in Determination of the Sustainability of Harvesting

One of the more starred findings of the review is the general lack of good quality data and the almost complete absence of a theoretical background for determining sustainable yield of NTFPs. Consequently, despite the many hopes that NTFPs can be managed sustainably, we do not actually have the data or the means to demonstrate that this is possible It is suggested that this is a very severe and urgent problem.

Application of Novel Sampling Strategies to NTFPs

There are a range of newer (at least to forestry) sampling techniques such as rank set sampling and adaptive cluster sampling, that appear to offer certain advantages for NTFPs. Clearly there is new methodology available, however, much of it may be at the cutting edge of statistics and has not yet been considered for NTFPs.

Cross-disciplinary Exchange of Ideas and Methods Suitable for use with NTFPs

There is an urgent need for cross-disciplinary exchange in the development of NTFP specific sampling designs and methods. There are many ideas and methods that apply in a wide range of different disciplines but there is little cross-over between them. In this respect the reviewer suggests we should think of NTFPs as wild-grown products from natural or seminatural environments. Although there are many products that lie in a grey area between the wild and domestic, it seems that the problems of managing cultivated or wild products are quite different and this needs to be reflected in methods selected for their quantification, management and monitoring. Areas where techniques and ideas can be shared between domestic and wild resources might also be identified.

Effective Communication of Advice to Field Workers and Communities

To be effective, any advice that can be offered and the results of any research undertaken has to get to people on the ground. Research must be developed in the field in a manner responsive to expressed needs and then put into a form appropriate for use by field technicians and people in village communities.

Possible Solutions for Difficulties in NTFP Inventory

In Nepal different organizations have developed different methods of NTFP inventory. Some have focused more on participatory approach while others emphasize on field measurement. These two methods should be combined to get the benefit of each other and exclude their demerits. Some of the possible strategies for NTFP inventory are;

- ✰ Inventory methods which are cost effective and locally adapted should be preferred.
- ✰ Indigenous and scientific knowledge should be linked to make inventory effective and more precise.

- ☆ Inventory methods for the hills and terai even for same species should be adopted accordingly.

Use of Aerial Photographs in Forestry Inventory

Remote Sensing

- ☆ Remote sensing is the science and art of obtaining information about an object, area or phenomenon through analysis of data acquired by a device that is not in contact with the object, area or phenomenon under investigation.
- ☆ Aerial remote sensing-it is that method of remote sensing in which cameras or other devices, fixed in an aircraft flying at fixed altitude are used to take photographs of any resource on earth. It is also called aerial photography.
- ☆ Space remote sensing- it is that method of remote sensing in which cameras, sensor or other devices attached to a satellite orbiting round the earth, take photographs of earth and resources inside it.

Aerial Photography

- ☆ Aerial photography is the process of taking photograph of earth's surface and its natural resources from an aircraft flying at fixed altitude from the earth
- ☆ Aerial photographs are the photographs of earth's surface and its natural resources taken from an aircraft flying at fixed altitude from the earth.
- ☆ The relative position of the objects on photographs is superficially similar to the actual position on ground.
- ☆ Canada was the first country which applied aerial photography in forestry.
- ☆ In Asia(tropics), Burma is the first country to start aerial survey of vegetation.
- ☆ In Nepal, Aerial photography was used for the estimation of forest and shrub land cover and changes.
- ☆ Aerial photography are very useful tool for the forest manager.
- ☆ A basic knowledge of the location and extent of the forest is critical to the management of forest resources.
- ☆ Aerial photographs are useful in designing and conducting field inventories and used to estimate tree and stand characteristics

- Scope
- To determine the volume (value) of standing trees in a given area.
- To get a reliable estimate of the forest area.
- The measurement of all or an unbiased sample of trees within the area
- To get the knowledge of location of all tract corners and boundary lines

Types of Aerial Photographs

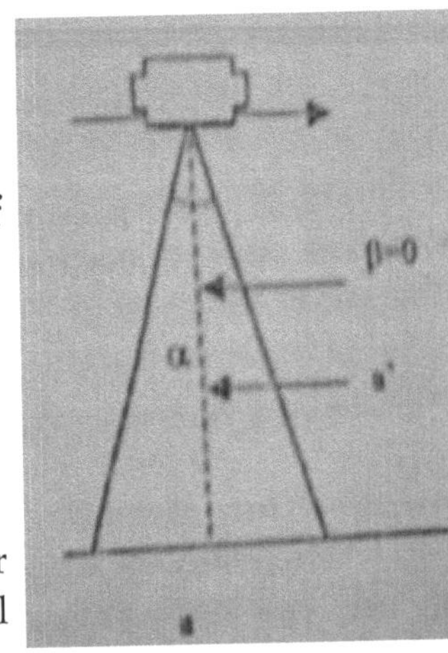

Aerial photography may be classified into various types depending on the different criteria.

On the basis of Position of Optical Axis of Cameras

- Vertical photograph
- Oblique photograph

Vertical Photograph

- Optical axis of camera is kept perpendicular or nearly perpendicular to the horizontal plane
- The degree of tilt is less than 4^0.
- It is considered to be best because ground features like building, roads, streams, forest boundaries appear same as the map of similar scale covers small area but scale is quite uniform over the whole picture.

  α= angle of view

  β= deviation of optical axis from vertical line

  a = vertical line perpendicular to the horizon

Oblique Photographs

Low Oblique Photographs

- Optical axis of camera is tilted by 30 degree or less from the vertical and horizon does not show in the picture

High Oblique Photographs

- Optical axis of camera is tilted by 60 degree and horizon is apparent.

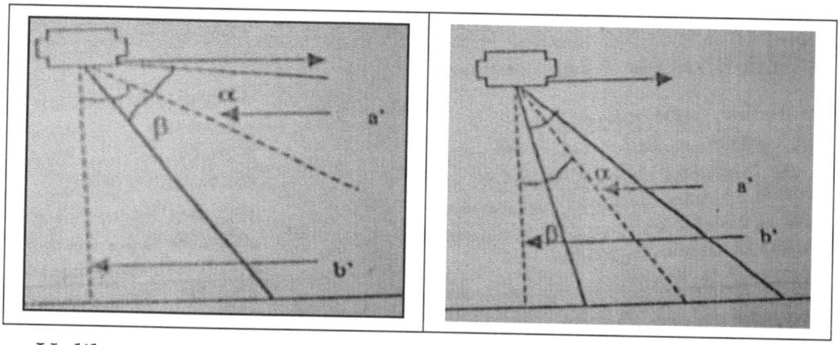

- ☆ Unlike vertical photograph the scale of oblique is variable, and that is why there is distortion.
- ☆ The degree of distortion increases towards the horizon

On the Basis of Films or Sensor used

- ☆ Panchromatic black and white photograph
- ☆ Infra-red black and white photograph
- ☆ Normal color photograph
- ☆ Infra-red color photograph

Panchromatic Black and White Photograph

- ☆ This film is sensitive to the electromagnetic spectrum in the wavelength range of 0.3 to 0.75 μm
- ☆ When panchromatic film is used, the conventional black and white photograph is produced.
- ☆ This film has low sensitivity in the green region of visible spectrum and therefore is not suitable for identification of plant species. Panchromatic photograph recording reflected sunlight over the wavelength band 0.4 to 0.7 μm

Infra-red Black and White Photograph

- ☆ By the use of infra-red film and proper filter, the electromagnetic spectrum in the wavelength range of 0.3 to 0.9 μm is photographed as black and white photograph.

- ☆ The photograph is very useful in identification of broadleaved and coniferous trees.

Normal Color Photograph

- ☆ A color film with proper filter produced colored photograph in true color. It registers visible color by human eye.
- ☆ It is three-layered film with emulsion sensitive to an additive primary color i.e. blue, green and red.
- ☆ The color photography is much more useful in separation and identification of different species and diseases.

Infra-red Color Photograph

- ☆ It is called false color photograph as it shows objects in different color as compared to the true color of the objects.
- ☆ The film sensitive to infrared region is used in taking such photographs.
- ☆ The normal color of vegetation is green but in infrared color photograph, vegetation appears red.

On the Basis of Device Used

- ☆ Single lens photograph-single lens camera used
- ☆ Multi-spectral photograph-more than one camera or camera with more than one lens used
- ☆ Multi-spectral imagery-optical mechanical scanner used

On the Basis of Scale of Photograph

- ☆ Small scale photograph-1.40,000 to 1.70,000 or more
- ☆ Medium scale photograph-1:20,000 to 1: 40,000
- ☆ Large scale photograph-1:5000 to 1:20,000

Season of Photography

- ☆ Depends upon the nature of the features to be identified, the film to be used and number of days suitable for photographic flights
- ☆ Less expensive in sunny area
- ☆ Best season-during October to February
- ☆ Another season-growing season
- ☆ Eg. Coniferous forests-October to November, Mixed deciduous forests-December to February, Tropical evergreen and moist deciduous forests-January to February
- ☆ Spring, Winter, Summer photographs

Scale of Photographs

- ☆ Scale= photo distance/ground distance or
- ☆ Scale= focal length of camera (f)/flying height above ground ('h')

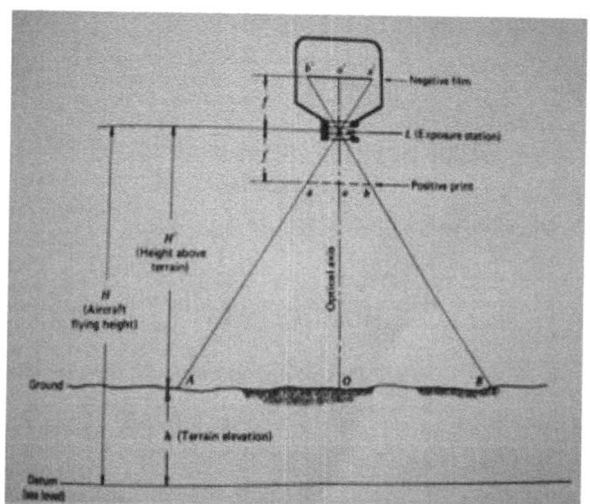

Figure 3.7 Scale of a vertical photograph taken over flat terrain.

b. Photo Interpretation & Technical Terms

- ☆ Photo interpretation is an act of examining photographic images and judging their significance.
- ☆ When we can identify what we see on the photographs and communicate this information to others, we are practicing aerial photographs interpretation.
- ☆ Aerial photo graphs contain raw photographic data. These data, when processed by a human interpreter's brain, become useful information.
- ☆ Aerial photographs can be interpreted by using following basic elements/characteristics: Shape, Size, Pattern, tone, texture, shadows, site, association and resolution.

Tone (or hue)

- ☆ It refers to the relative brightness or color of objects on an image.
- ☆ Relative photo tones could be used to distinguish between deciduous (light tone) and coniferous trees (dark tone) on black and white photographs
- ☆ Without tonal difference, the shape, patterns, and texture of object could not be distinguished.

Shape

- ☆ It refers to the general form, configuration or outline of individual objects.
- ☆ Shape help a great deal in making interpretation in forestry.
- ☆ Eg. Some species have rounded crowns, some have cone-shaped crowns and some have star shaped crowns.
- ☆ Variation of these basic crown shapes also occur.

Pattern

- ☆ It related to the spatial arrangement of objects.
- ☆ The repetition of certain general forms or relationship is characteristic of many objects, both natural and constructed and gives objects a pattern that aids the image interpreter in recognizing them.
- ☆ For example, the ordered spatial arrangements of tree is an orchard, is in distinct contrast to that of forest tree stands.

Size

- ☆ Size of objects on images must be considered in the context of image scale. A small storage shed, for example might be misinterpreted as barn if size were not considered.
- ☆ Relative sizes among objects on images of the same scale must also be considered.
- ☆ Similarly, size is also helpful in identifying the forest stands.

Texture

- ☆ It is the frequency of tonal change on a photo.
- ☆ Texture is produced by an aggregation of unit features that may be too small to be discerned individually on the image, such as tree leaves and leaf shadows.

- It is the product of their individual shape, size, pattern, shadow.
- It determines the overall visual "smoothness" or "coarseness" of image features.
- Difference in texture help distinguish tree species.

Shadow

- Shadows are important to interpreters in two opposite respects: 1). the shape of outline of a shadow affords and impression of the profile view of objects (which aids interpretation) and 2) Objects within shadows reflect little sight and are difficult to discern on a image (which hinders the interpretation)
- When trees are isolated, shadows provide a profile of image of trees that is useful in species identification.

Sites

- It refers to topographic or geographic location and is a particularly important in identification of vegetation types.
- For eg., Certain tree species would be expected to occur on well drained upland sites, whereas other tree species would be expected to occur on poorly drained lowland sites.
- *Alnus nepalensis* is expected to occur in degraded and exposed sites.

Association

- It refers to the occurrence of certain features in relation to others.
- Eg. Stalk of wood nearby the forest would be easier to identify than the stack of wood kept in proximity with small shed in the agriculture land.

Resolution

- It depends on many factors, but it always places a practical limit on interpretation because some objects are too small or have too little contrast with their surroundings to be clearly seen on the image.
- There are certain equipments that can be used for the aerial photo interpretation. They help in viewing the photo, making measurement on the photo, performing interpretation tasks and transferring interpreted information to base maps or database.

c. Forest Classification

- ✰ Forest classification can be done on aerial photographs using the basic elements for photo interpretation.
- ✰ The classification of forest types and plant species largely depends on quality, scale and season of photography and the ability of interpreter.
- ✰ The shape, texture, and tone or tree foliage as seen on vertical photographic along with other basic elements of photo interpretation are keys to forest classification.
- ✰ The first step in forest classification is to determine which types should and should not be expected in given locality.
- ✰ It will also be helpful to get familiar with the most common plant and environmental association.
- ✰ This information can be derived by previous maps or direct observation.
- ✰ In some regions, photo-interpretation keys are available for recognition of forest cover types and such keys can be used.
- ✰ Classification can be conducted in minute details in large scale photographs where species can be identified and in small scale photographs, the knowledge of locality is essential for forest classification.
- ✰ In modern days, computer programs (image processing software like ERDAS, PCI, ENVI, GRASS etc) are available for making forest classification using the aerial photographs and other imageries.

Relationship between photographic scale and expected levels of plant recognition

Photographic Scale	*General level of plant discrimination*
1:30,000-1:100,000	Recognition of broad vegetative types, largely by inferential processes
1: 10,000-1:30,000	Direct identification of major cover types of species occurring in pure stands
1:2,500-1:10,000	Identification of individual trees and large shrubs
1:500-1:2,500	Identification of individual range plant are grassland types

d. Area Determination

- ✰ A truly vertical aerial photo is just like a map and area can be measured with methods similar to that from a map.
- ✰ The scale of photographs is known and the area is calculated accordingly.

- ☆ But in aerial photo, the center part of the plant is generally true as map but as we move radially outward from the center, the scale changes and there is displacement.
- ☆ In determining the area, if all corners are at the same elevation, the area calculation from photographs can be made accurately. If not, substantial error can exist.
- ☆ Similarly, if the area are measured outward the center of the photograph, then some correction will only yield accurate result.
- ☆ Thus, we should use the photo where the area to be measured is as close to nadir as possible. Such errors can be reduced by using photos taken from higher flying height.
- ☆ Following methods and tools are used to measure area in aerial photographs- Dot grids, planimeter, weight apportionment and GIS method.

Dot Grids

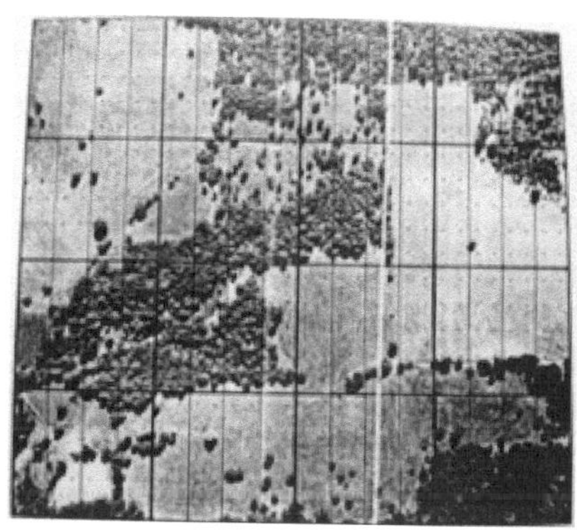

- ☆ Determine number of dots per unit area
- ☆ Count number of dots in stand or feature on map or photo
- ☆ Calculate the area of stand or feature
- ☆ Convert measurement to ground areas
- ☆ Widely used method of area calculation on aerial photo
- ☆ Relatively simple and inexpensive tool for estimating area on photographs

Planimeter

- ☆ Run the pointer of an instrument around the boundaries of an area in clockwise direction; usually the perimeter is traced two or three times for an average and read the instrument details.
- ☆ Convert measurements to ground areas
- ☆ Laborious if many areas need to measure

Weight Apportionment

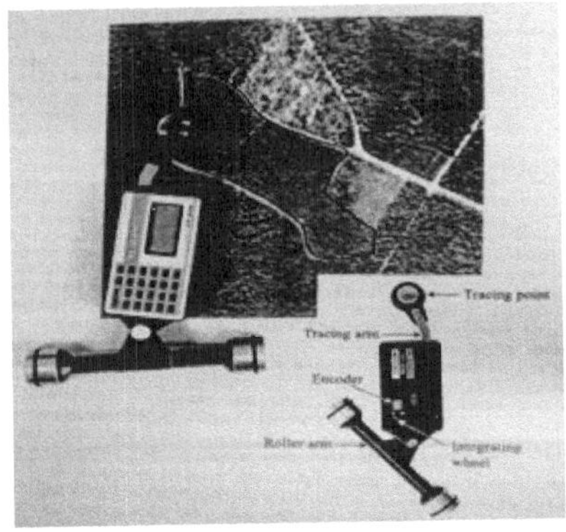

- ☆ Traditional method
- ☆ Physically cut a method into individual areas
- ☆ Weigh each area and determine areas based on their weight
- ☆ Need a very sensitive instrument to weight small pieces of paper

GIS Method

- ☆ Advance method
- ☆ By digitizing the particular area and using the different function, area can be calculated

Volume Estimation

- ☆ For estimating individual tree volume, multiple entry tree volume tables based on dbh and height can be converted to aerial volume table when correlations can be established between crown diameter and stem diameters
- ☆ Photo determination of crown diameter is substituted for the usual ground measures of dbh, and total heights are measured on stereoscopic pairs of photographs.
- ☆ Large scale aerial photographs are essential for obtaining reliable crown diameter and health measurement of individual trees.
- ☆ Where only small-scale aerial photographs are available, stand variables should be measured than individual tree variable
- ☆ Aerial stand volume tables are multiple entry tables that are used based on assessments of two or three photographic characteristics of the dominant –co dominant crown canopy; average stand height, average crown diameter and percent of crown close
- ☆ These tables are usually derived by multiple regression analysis; photographic measurements of independent variables are made by several skilled interpreters when developing the volume prediction equation.

Other Forestry Applications of Aerial Photography:

- ☆ Vegetation growth distribution investigation
- ☆ Forest resource investigation
- ☆ Forest fire monitoring
- ☆ Forest disease and pest monitoring
- ☆ Shifting cultivation study
- ☆ Timber harvest planning
- ☆ Monitoring of logging and reforestation
- ☆ Forest recreation resource inventory and monitoring
- ☆ Wildlife habitat analysis
- ☆ Planning forest roads
- ☆ Monitoring power line right of way vegetation growth etc.

UNIT 5

Objective Questions with Answers

1. **Mensuration is derived from __________**

   (a) Russia (b) Latin

   (c) Greek (d) (a) & (b)

2. **One hectare ____________**

   (a) 2.47 acres (b) 2.54 acres

   (c) 2.49 acres (d) 2.59 acres

3. **In metric system girth of a tree is always taken in**

   (a) Metres and to the nearest centimeter.

   (b) Foot and to the nearest inches.

   (c) Centimeters to the nearest millimeter.

   (d) Only in foot.

4. **The FPS systems enacted during the year**

   (a) 1956 (b) 1966

   (c) 1962 (d) 1967

5. **The CGS or metric systems enacted during the year**

   (a) 1956 (b) 1966

   (c) 1962 (d) 1967

6. **The diameter class (cm) followed in India**
   (a) 1,2,5 (b) 2,5,10
   (c) 1,3,5, (d) 3,5,10
7. **The diameter class (cm) followed in India**
   (a) 1,2,5 (b) 2,5,10
   (c) 5,15,30 (d) 3,5,10
8. **The height class (m) following in India are**
   (a) 1,3,5 (b) 1,2,5
   (c) 1,3,10 (d) 1,2,10
9 **Height class followed for tree attaining a height of 25 m**
   (a) 5 (b) 10
   (c) 2 (d) 1
10. **Which of the following is not correctly matched**
   (a) Mana – Measure of volume
   (b) Tulamana – Measure of weight
   (c) Avamana – Measure of diameter
   (d) Kalapramana – Measure of time
11. **The country, that adopts 1.3 m as breast height**
   (a) Burma (b) America
   (c) Europe (d) India
12. **Basal Area of a tree refers to**
   (a) Cross sectional area at stem height
   (b) Cross sectional area at base
   (c) Cross sectional area at top height
   (d) Cross sectional area at breast height
13. **On sloping ground, the dbh should be measured on**
   (a) Down hill side (b) Accessible side
   (c) Up hill side (d) None of these
14. **The maximum spread of the crown along its widest diameter is**
   (a) Crown length (b) Crown height
   (c) Crown width (d) Crown diameter

15. Relationship between the diameter and girth of trees is

(a) ½ (b) 0.3182

(c) 7/22 (d) All the above

16. Volume of the log calculated using the formula

(a) g/4x1 (b) (g2/4) x1

(c) (g/4)2x1 (d) None

17. The instrument designed based on the principal of similar triangle

(a) Christen Hypsometer (b) Brandis Hypsometer

(c) Altimeter (d) Abney level

18. Instrument used for measuring the height of the tree based on trigonometrical principles.

(a) Christen's hypsometer (b) Smithies hypsometer

(c) Improvised calipers (d) Brandis hypsometer.

19. Abney's level is used for

(a) Measurement of height

(b) Measurement of volume

(c) Measurements of hight and contour survey

(d) Past rate of growth of a tree

20. Instrument used for measure the basal area of the crop

(a) Relaskop (b) Abney's level

(c) Brandis hypsometer (d) Improvised caliper

21. Pick the Newton's formula

(a) (S1+4Sm+S2/6)x1 (b) Smx1

(c) S1 (d) S1+S2/2x1

22. Factors affecting the overall weight of a quantity of wood are

(a) Density and moisture content

(b) Bank and foreign material

(c) Moisture content, Bank and foreign material

(d) All of these

23. Pressler's increment borer is used to determine

(a) Increment of a tree (b) Height of a tree

(c) Volume of a tree (d) Age of a tree

24. Metzer's theory is related to

(a) Stem form
(b) Form height
(c) Sampling
(d) None of these

25. In slopping ground diameter of tree is measured on

(a) Down hill side
(b) Uphill side
(c) Leaning side
(d) None of these

26. Stump analysis is done to determine

(a) Age and diameter relation
(b) Age and Height relation
(c) Age and Volume relation
(d) All of these

27. The method of sampling adopted in Bamboo enumeration is

(a) Multistage sampling
(b) Multiphase sampling
(c) Random sampling
(d) List sampling

28. The breast height adopted by most of the European countries is

(a) 1.3m
(b) 1.37m
(c) 1.27m
(d) 3.0m

29. The normal size of the sample plot is

(a) 0.1 to 0.2 ha
(b) 0.1 to 0.5 ha
(c) 0.1 to 1.0 ha
(d) 0.1 to 2.0 ha

30. The Instrument, which is used to measure angle of slope, is

(a) Altimeter
(b) Clinometers
(c) Hypsometer
(d) Abney's level

31. Artificial form factor is also known as

(a) Absolute form factor
(b) Breast height form factor
(c) Normal form factor
(d) Tree form factor

32. True form factor is also known as

(a) Absolute form factor
(b) Breast height form factor
(c) Normal form factor
(d) Tree form factor

33. Crop diameter refers to

(a) Mean diameter of tree in a stand

(b) Mean diameter of Dominant trees

(c) Mean diameter of suppressed trees

(d) None of these

34. The standard of weights and measures act enacted during the year

(a) 1956 (b) 1966

(c) 1957 (d) 1967

35. The product of form factor and total height of tree

(a) Form quotient (b) Form class

(c) Tree Height (d) Form Height

36 Crop diameter is

(a) Mean diameter of tree in a stand

(b) Diameter corresponding to the mean basal area of uniform crop

(c) Mean diameter of dominant trees

(d) None of these

37. Basal area of tree implies

(a) Cross sectional area at stump height

(b) Cross sectional area at breast height

(c) Cross sectional area at base.

(d) None of these

38. Age of trees with annual rings determined by

(a) Swedish bark gauge (b) Wheeler penta prism

(c) Pressler's increment borer (d) Blume-leiss hypsometer

39. Average annual increment for any short period is called as

(a) CAI (b) PAI

(c) MAI (d) IP

40. The relationship between CAI/MAI (Pick the odd out)

(a) CAI is greater than MAI, the MAI is raising

(b) To begin MAI keeps below CAI

(c) CAI attains its maximum before MAI

(d) MAI is fully under CAI

41. Which instrument is used to determine the age of a tree

(a) Abney's level (b) Pressler's borer

(c) Relascope (d) Altimeter

42. Effective variables for volume table preparation

(a) Single variable (b) Double variable

(c) Trible variable (d) Multi variable

43. The volume tables are combined fom DBH alone is called

(a) General volume tables (b) Regional volume tables

(c) Local volume tables (d) Standard volume tables.

44. Crop height is determined by

(a) Andre's formula (b) Heyer's formula

(c) Gumpel's formula (d) Lorey's formula

45. Proportionate distribution of trees in selection forests is attributable to

(a) Mayers law (b) Faustman law

(c) De Liocourtslaw (d) None of these

46. Optimal number of sample plots for each thinning regime in yield table preparation is

(a) 500 (b) 600

(c) 700 (d) 800

47. The formula for converting d.o.b. into d.u.b. is

(a) G=g'-2…t (b) G'=g-2…h

(c) G=d'-2…t (d) G=g'-2…h

48. Smalian's formula for calculating volume of log is

(a) SI (b) Smx1

(c) S1+S2/2x1 (d) S1+S2…./ S1S2/3x1

49. Calculate the basal area per ha if the number of tallied trees was 14 and the BAF of the wedge prism used is three

(a) 42m^2 (b) 14 m^2

(c) 21 m^2 (d) None of these

50. Normal form quotient is defined as:-

(a) Mid diameter / d.b.h.

(b) Diameter at one half its height above breast height/ d.b.h.

(c) Diameter at any point/d.b.h.

(d) Mid diameter x d.b.h.

51. Who coined the term form quotient?

(a) Schiffel (b) Tor-Jonson

(c) Schneider (d) Pressler

52. In Normal Form factor basal area is measured at:-

(a) Breast height

(b) Any Convenient height

(c) A constant proportion of the total height

(d) None of these

53. Who proposed the theory to explain variations in taper from tree to tree

(a) Metzger (b) Huber

(c) Newton (d) Smalian .

54. The instrument 'wedge prism' is mainly used in

(a) Plot sampling (b) Horizontal point sampling

(c) Vertical point sampling (d) All of the above

55. One diopter is approximately equivalent to an angle of

(a) 0.57° (b) 0.47°

(c) 0.75° (d) 0.74°

56. Linear growth model is preferred only for

(a) Nursery stage (b) Sapling stage

(c) Pole stage (d) Tree stage

57. The standard length of a log expect the last section is

(a) 5m (b) 4m

(c) 3m (d) 6m

58. If the canopy density is between 0.75 and 1.0, then the forest is referred as

(a) Closed (b) Dense

(c) Thin (d) Open

59. Yield tables in which data are given for different grades of thinning

(a) Volume yield table (b) Money yield table

(c) Single yield table (d) Multiple yield tables

60. Yield table is generally prepared for

(a) Regular forest (b) Irregular forest

(c) Reserved forest (d) Mixed forest

61. The quarter girth formula to calculate volume of log is

(a) g21/4 (b) 4/g2x1

(c) g21/16 (d) 16/g2x1

62. What are the variables require enough for general volume table preparation?

(a) Diameter only (b) Height only

(c) Diameter and height (d) Diameter, height and form

63. Sampling units are taken at two or more stages and sampling units gets smaller at each successive stage of sampling is called as

(a) Unrestricted sampling (b) Stratified sampling

(c) Multistage sampling (d) Multiphase sampling

64. The height corresponding to the mean diameter of 250 biggest diameters per hectare as read from Diameter height curve is

(a) Top height (b) Crop height

(c) Mean height (d) Mean basal area

65. The age diameter, age-height, age-volume relations throughout the life of the trees are analyzed by

(a) Stump analysis (b) Stem analysis

(c) Pressler's increment borer (d) Mathematical formula

66. The instrument, which is used to measure diameter at any convenient height is

(a) Tree caliper (b) Tailor's tape

(c) Wheeler's Penta Prism (d) Relaskop

67. Volume of stem from ground level down to 20 cm diameter over bark is

(a) Commercial volume (b) Standard stem small wood

(c) Standard stem timber (d) Sawn out turn

68. In point sampling, P.P.S. is referred to as

(a) Poly areal plot sampling (b) Probability point sampling

(c) Probability plot sampling (d) Probability proportional to size

69. Normally the Fig. s of fractional quality running between the quality classes

(a) 0.0 to 1.0 (b) 0.0 to 10

(c) 0.0 to 0.2 (d) 0.0 to 20

70. Which one is not considered as yield table

(a) Volume yield table (b) Volume table

(c) Money yield table (d) Stand table

71. It is a measure of the relative productive capacity of a site for a particular species is called as

(a) Site quality (b) Site index

(c) Canopy density (d) Crown density

72. An instrument, which is particularly used for vertical point sampling is

(a) Wedge prism (b) Spiegal relaskip

(c) Telerelskop (d) Conimcter

73. Generally temporary sample plots are used for

(a) Enumeration survey

(b) Preparation of yield table

(c) Measurement of crop volume and increment

(d) Comparison of growth in different qualities of locality

74. Schneider's formula is used for determining

(a) Height of the tree (b) Growth of the tree

(c) Volume of the tree (d) Increment percentage

75. Which sampling method is random sampling method

(a) Selective sampling (b) Systematic sampling

(c) Stratified sampling (d) Sequential sampling

76. When the tree is forked above the breast height, it is counted as,

(a) One tree

(b) Each fork treated as separate tree

(c) Tree counted as one or two depend upon place measurement

(d) Cannot be determined

77. Site index is the relation between

(a) Age and site (b) Age and diameter

(c) Age and height (d) Age and volume

78. The DBH was measured in United Kingdom is at

(a) 1.37m (b) 1.30m

(c) 1.39m (d) 1.35m

79. On sloping ground the DBH is measured

(a) Down hill side (b) Uphill side

(c) Both up and down hill side (d) None of the above

80. The Fig. s obtained by Remote sensing visualizes facts by

(a) Orographic view (b) Orthographic view

(c) Perspective view (d) Perpendicular projection

81. Normal form factor also called

(a) True form factor (b) Absolute form factor

(c) Artificial form factor (d) Breast height form factor

82. In bamboo forest, the system of sampling adopted was

(a) Multistage sampling (b) Multiphase sampling

(c) Sequential sampling (d) Random sampling

83. Solid volume of fire wood is measured by

(a) Smalians formula (b) Hubers formula

(c) Hunden hagen formula (d) None of the above

84. Bole surface area calculated with the following parameters

(a) Girth and length of log

(b) Length and Breadth of log

(c) Girth and length with form substitutes

(d) (a) and (b)

85. Horizontal projection on the ground of the tree crown is called

(a) Crown cover (b) Crown width
(c) Crown length (d) Tree width

86 Vertical point sampling is attributable to

(a) Bitterlich (b) Hirata
(c) Avery (d) Chaturvedi

87. Who introduced the horizontal point sampling technique?

(a) Hirata (b) Hirata
(c) Bitterlich (d) Smalin

88. The instrument 'wedge prism' mainly used in

(a) Plot sampling (b) Horizontal point sampling
(c) Vertical point sampling (d) All of the above

89. If the canopy density is between 0.75 and 1.0, t hen the forest is referred as

(a) Closed (b) Dense
(c) Thin (d) Open

90 The branch of forestry which deals with the determination of dimension, from and age and increment of single trees, stands or whole woods either standing or after felling is called

(a) Dendrology (b) Forest Management
(c) Dendrometry (d) Forest Protection

91. What are the variables require enough for general volume table preparation?

(a) Diameter only (b) Height only
(c) Diameter and height (d) Diameter, height and form

92. Sampling units are taken at two or more stages and sampling units gets smaller at each successive stage of sampling is called as

(a) Unrestricted sampling (b) Stratified sampling
(c) Multistage sampling (d) Multiphase sampling

93. If the focal length of the camera is 20 cm and the flying height above the ground level is 5000 m, calculate the photo scale

(a) 1:10000 (b) 1:15000
(c) 1:20000 (d) 1:25000

94. The following are some clinomete5rs used in leveling. Pick the odd man out.

(a) Ghat tracer (b) Topographic Abney level

(c) De Lisle clinometer (d) Tangent clinometers

95. Hyposometric leveling is a method to find the

(a) Height of mountains (b) Depth of water bodies

(c) Altitudes (d) Height of trees

96. Importance of Value Index (IVI) is used for

(a) Estimation of biomass (b) Mapping of forests

(c) Phyto-sociology of forests (d) Survey of vegetation

97. Definition of 'Productivity' is

(a) Biomass produced in one ha

(b) Biomass produced in one acre

(c) Biomass produced in one ha per year

(d) Biomass produced after one rotation

98. Volume of stem wood in round between 20 cm and 5 cm diameter over bark is called

(a) Commercial volume (b) Standard stem small wood

(c) Standard stem timber (d) Sawn out turn

99. Volume of stem from ground level down to thin end of diameter over bark is

(a) Commercial volume (b) Standard stem small wood

(c) Standard stem timber (d) Sawn out turn

100. The Latin name for Andaman Padauk tree is

(a) Veteria indica (b) Pterocarpus dalbergioides

(c) Shorea robusta (d) Dalbergia latifolia

101. The butt end of a tree resembles frustum of

(a) Trapezium (b) Paraboloid

(c) Cone (d) Neiloid

102. Top portion of tree resembles frustum of

(a) Neiloid (b) Paraboloid

(c) Cone (d) Trapezium

103. Middle portion of tree resembles frustum of

(a) Neiloid (b) Paraboloid

(c) Cone (d) Trapezium

104. Measure of the value of the angle expressed as its Sine is termed as

(a) Diopter (b) BAF

(c) Calibration distance factor (d) Plot radius factor

105. The sample plot with minimum perimeter for a given area is

(a) Square (b) Rectangular

(c) Triangular (d) Circular

106. Difficulty in applying yield tables to natural forests is

(a) Lack of a representative age (b) Uniform site quality

(c) Over stocking (d) Over matured crop

107. Spectral signature of green vegetation in a satellite captured image is

(a) White (b) Green

(c) Blue (d) Red

108. Vertical point sampling is used for estimating

(a) BA/Ha (b) No of stems/Ha

(c) Mean stand height (d) All of the above

109 Panchromatic camera of IKNOS staellite has the highest ground resolution of

(a) 1 m (b) 1.5m

(c) 5.8 m (d) 8.5 m

110. Stump analysis is used to assess the relation of

(a) Age and growth (b) Diameter and volume

(c) Height and diameter (d) Height, diameter and volume

111. Tree caliper will give

(a) Accurate measurement (b) Consistence measurement

(c) Absolute measurement (d) None of these

112. Christens hypsometer is used to measure the tree

(a) Volume (b) Taper

(c) Height (d) None of these

113. Which of the following is correctly matched

(a) Clinometers-density (b) Altimeter-atmospheric pressure

(c) Hypsometer-height (d) Relaskop-light intensity

114. Indian Institute of Remote Sensing is located at

(a) Dehradun (b) Bangalore

(c) New Delhi (d) Chandigarh

115. The ideal time for taking Aerial photograph is

(a) 9 to 11 am (b) 11 to 12 pm

(c) 2 to 3 pm (d) 3 to 4 pm

116. Metzgeris Theory or Girder theory discusses

(a) Tree stem form (b) Tree volume

(c) Tree height (d) Species density

117. Horizontal Point is used for determination of

(a) Basal area for hectare (b) Number of stems per hectare

(c) Volume per hectare (d) All of the above

118. Stem analysis is done to determine

(a) Age of a tree (b) Height of a tree

(c) Volume of a tree (d) Past rate of growth of a tree

119. The rotation at which MAI is maximum

(a) Rotation of height income

(b) Financial rotation

(c) Rotation of maximum volume production

(d) All of these

120. In remote sensing, the scale for small scale photography ranges between

(a) 1:40000 and 1:70000 (b) 1:20000 and 1:40000

(c) 1:5000 and 1:20000 (d) None of these

121. The Forest Survey of India (FSI) conducts Forest Cover Survey once in every

(a) 2 years (b) 5 years

(c) 10 years (d) 4 years

122. Small scale photograph has scale of

(a) 1:40000 and 1:70000
(b) 1:20000 and 1:40000
(c) 1:5000 and 1:20000
(d) 1:50000 to 1:100000

123. The productive capacity of Indian Forest is

(a) 2.1 m^3/ha
(b) 0.7 m^3/ha
(c) 1 m^3/ha
(d) 0.5 m^3/ha

124. The instrument not used in horizontal point sampling is

(a) Simple angle gauge
(b) Dumpy level
(c) Spiegel releskope
(d) Wedge prism

125. Form factor is worked out using the formula

(a) F=V/Sh
(b) F=Sh/V
(c) F=V/S
(d) F=V/h

126. The diameter corresponding to 250 biggest diameters per hectare is called

(a) Crop diameter
(b) Mean diameter
(c) Top diameter
(d) Average diameter

127. The term normal form quotient is attributable to

(a) Tor Jonson
(b) Schiffel
(c) Hojer
(d) Behre

128. Instrument used measuring the height of the tree based on trigonometical principles

(a) Christen's hypsometer
(b) Smithies hypsmeter
(c) Improvised calipers
(d) Brandis hypsometer

129. Wooden scale is used to measure

(a) Diameter of standing tree
(b) Diameter of felled tree
(c) Girth of standing tree
(d) Girth of felled tree

130. Which instrument is used to measure diameter at any height of a tree?

(a) Abney's level
(b) Christen's hypsometer
(c) Tree caliper
(d) Wheeler's pentaprism

131. Solid volume of the stack can be determined by

(a) Timber calculators
(b) Xylometric method
(c) Specific gravity method
(d) None of these

132. Local volume table are applicable to

(a) Working circle (b) Block

(c) Compartment (d) Range

133. Girth is equal to

(a) ….d2 (b) …d

(c) 2… (d) …d/4

134. The error involved in using caliper is

(a) Positive (b) Negative

(c) Both (a) and (b) (d) None of these

135. Give the formula for basal area

(a) Ah/2 (b) …$d^2/2$

(c) …$d^2/4$ (d) …ab^2

136. Horizontal point is use for determination of

(a) Volume/ha (b) Basal area/ha

(c) Height (d) None of these

137. Random sampling (tick the odd man)

(a) Simple (b) Stratified

(c) Multistage (d) Systematic

138. Non random sampling

(a) Systematic (b) Selective

(c) Sequential (d) List

139. A table showing for a given species average weight of trees or logs for one or more dimensions like girth, height or DBH is termed as

(a) Stand table (b) Money table

(c) Biomass table (d) Yield table

140. The limit of standard small wood

(a) Down to 20 cm diameter over bark

(b) Down to 20 inch diameter under bark

(c) Down to 5 cm diameter under bark

(d) Down to 5 cm diameter over bark

141. A table showing the distribution of stems by diameter classes is referred as

(a) Stand table (b) Local volume table

(c) Yield table (d) None of these

142. The abney level is used to measure ....

(a) Tree height (b) Dbh

(c) Board feet (d) All of the above

143. Calipers are used to measure ....

(a) DBH (b) Tree height

(c) Board feet (d) All of the above

144. A tool used to help read aerial photographs is the ....

(a) Haga altimeter (b) Clinometers

(c) Stereoscope (d) Abney level

145. Which tool is used for timber cruising?

(a) Wedge prism (b) Increment borer

(c) Cant hook (d) Chain saw

146. The haga altimeter will measure. ....

(a) Tree height (b) Dbh

(c) Basal area (d) Board feet

147. A log of marketable size which is made unmarketable because of rot, defect or poor form is called a ....

(a) Slash (b) Cull

(c) Scratch (d) Defects timber

148. A log rule is a table or formula used to determine tree:

(a) Age (b) Diameter

(c) Height (d) Volume

149. A bark gauge measures. ....

(a) Diameter of the tree trunk

(b) Ratio of bark to wood

(c) The decibels generated when a felled hits the ground

(d) Tree bark thickness

150 A wedge prism is used in a type of inventory called:

(a) Cluster sampling (b) Double sampling

(c) Plot sampling (d) Point sampling

151. On a topographic map which of the following is a dark, heavy line with an elevation Fig. ?

(a) Contour index (b) Contour interval

(c) Contour line (d) Horizontal control station

152. The Biltmore stick is used to measure. ....

(a) Tree height (b) Dbh

(c) Board feet (d) Land area

153. The area of the end of a log is called the .... area .

(a) Basal (b) End

(c) Stump (d) Butt

154. To perform a basal area sample, a forester would use which of the following tools:

(a) Diameter tape (b) Prism

(c) Realkascope (d) All of the above

155. The diameter of trees is usually measured at:

(a) Breast height (b) 4 and ½ feet above the ground

(c) Dbh (d) All of the above

156. Which of the following is/are an advantage of prism?

(a) Fast (b) Simple to use properly

(c) Reasonable accurate (d) All of the above

157. DBH can be measured by all of the following tools except ....

(a) Diameter tape (b) Biltmore stick

(c) Merit hypsometer (d) Calipers

158. Which of the following will NOT measure the height of a tree?

(a) Merit hypsometer (b) Abney level

(c) Clinometers (d) Log scale stick

159. Diameter measurements for sawlogs are taken at :

(a) The large end of the log (b) The log center

(c) The small end of the log (d) The stump height mark

160. When using the prism, a tree to be counted should be:

(a) All trees in the plot regardless of offset

(b) Offset completely

(c) Partially offset

(d) All of the above

161. The general term used to describe instruments used to measure the diameter of a tree is:

(a) Altimeter (b) Clinometers

(c) Dendrometer (d) Hypsometer

162. A hand held instrument used for measuring tree heights, slopes and vertical angles is a/an ....

(a) Abney level (b) Clinometers

(c) Stereoscope (d) Merritt hypsometer

163. A wedge prism is used to measure the .... of trees in a stand .

(a) Basal area (b) Board foot volume

(c) Diameter (d) Cubic foot volume

164. When performing a forest inventory with a wedge prism you are selecting trees by :

(a) Species (b) Their dbh

(c) Their distance form you (d) Using a table of random numbers to identify the trees

165. An increment borer is NOT used to do which of the following?

(a) Estimate the age of a tree (b) Determine the growth rate of a tree

(c) Eliminate knots in lumber (d) An increment borer does all of the above

166. Tree diameter growth is related to ....

(a) Age of the tree (b) Availability of moisture and nutrients

(c) Stand density (d) All of the above

167. A .... is used to estimate basal area per acre.

(a) Clinometers (b) Hypsometer

(c) Planimeter (d) Wedge prism

168. A Jacob staff is used with a .... to increase accuracy of tree measurement readings.

(a) Diameter tape (b) Planimeter

(c) Relaskop (d) Stereoscope

169. How many square feet are there in an acre?

(a) 40,000 (b) 43,522

(c) 43,560 (d) 10,000

170. Which of the following is the most limiting factor in the growth and development of an individual tree?

(a) Nutrients (b) Sunlight

(c) Rainfall (d) Temperature

171. Foresters recognize which of the following as the basic management unit in a forest?

(a) Stand (b) Lot

(c) Section (d) None of the above

172. DBH is :

(a) Diameter at breast height

(b) 4.5 feet above the ground

(c) A standard accepted in forestry Worldwide

(d) All of the above

173. General volume tab les which are applicable to large geographical area are based on

(a) One variable (b) Two variable

(c) Three variable (d) Four variable

174. Which of the following tools would NOT be used to determine tree height?

(a) Clinometers (b) Abney level

(c) Hypsometer (d) None of these

175. Which is the closest to a hectare?

(a) 1 acre (b) 2.5 acres

(c) 5.2 acres (d) 7 acres

176. The radius of a one-fifth acre plot is :

(a) 52.7 feet (b) 37.2 feet

(c) 27.5 feet (d) 73.2 feet

177. Of the tools for measuring DBH, which is the most precise?

(a) Caliper (b) Biltmore stick

(c) Diameter tape (d) Carpenter's rule

178. The scaling diameter for saw logs is normally measured at :

(a) The big end inside bark (b) The big end outside bark

(c) The small end inside bark (d) The small end outside bark

179. The term stumpage refers to :

(a) Stumps left after harvesting

(b) The diameter of trees measured 1 foot off the ground

(c) The value of trees standing in a woodlot

(d) A measure of tree volume

180. Which of the following instruments measures angles as well as direction?

(a) Clinometers (b) Abney level

(c) Realkascope (d) None of the above

181. An abandoned field changes into a mature forest over an extended period of time through a process called:

(a) Succession (b) Dominance

(c) Recycling (d) All of the above

182. Management of forests for multiple uses involves which of the following uses:

(a) Timber (b) Recreation

(c) Water (d) All of the above

183. A log rule is a table or formula used to determine tree:

(a) Age (b) Diameter

(c) Height (d) Volume

184. Basal area of a woodlot is given in what units?

(a) Square feet/acre (b) Cubic feet/acre

(c) Board feet/acre (d) Number of trees/acre

185. Merchantable height is :

(a) Commonly measured with a log scale

(b) Measured at the first fork in the tree depending on the species

(c) Measured to the minimum diameter based on products to be taken

(d) Not normally considered in the volume table

(e) Both (b) and (c)

186. The number of stems falls off from one diameter class to the next in geometrical progression, which means that the percentage reduction in the stem number from one diameter class to the next is constant. This is referred to as

(a) Brasnett's Law (b) Knuchel's Law

(c) De Liocourt's Law (d) Osmaston's Law

187. Tick the odd man out among the following technical terms connected with increfment

(a) Mean Annual increment (b) Increment percent

(c) Current annual percent (d) Terminal annual increment

188. If girth over bark (g) and bark thickness (t) are respectively 30 cm and 2 cm find out the girth under bark (g). Assume? To be 3.0

(a) 6cm (b) 12 cm

(c) 18 cm (d) 36 cm

189. One of the following instruments used for tree height measurement is not based on trigonometric principle. Identify the same.

(a) Brandis hypsometer (b) Smythie's hypsometer

(c) Blume Leiss hypsometer (d) Abney's level

190. The following are some methods developed for volume measurement in the sample plot. Identify the odd man out.

(a) Huber's method (b) Urich's method

(c) Gumple's method (d) Hartig's method

191. The product of form factor and total height of tree is defined as

(a) Form quotient (b) Farm class

(c) Form height (d) Form point ratio

192. Which one of the following is not a formula for tree form?

(a) Behre's formula (b) Hojer's formula

(c) Metzger's formula (d) Smalian's formula

193. Volume table based on diameter, height and form quotient are called as

(a) Form class volume table (b) General volume table

(c) Standard volume table (d) Commercial volume table

194. The following are some of the terms for point sampling. Identify the odd man out

(a) Angle count cruising (b) Pointless cruising

(c) Variable plot cruising (d) Polyplot cruising

195. The following are some kinds of volume tables. Identify the odd man out

(a) Multiple yield table (b) Variable yield table

(c) Single yield table (d) Money yield table

196 Crown length is vertical measurement of the crown of a tree from the tip of the crown of a tree to the

(a) Lowest green branch on the bole

(b) Lowest green branch forming green crown all around

(c) Lowest branch green or dead

(d) Pont halfway between the lowest green branch forming green crown all around and the lowest green branch on the bole

197. Which height is used for determine the volume of a tree crop?

(a) Mean height (b) Mean height or crop height

(c) Top height (d) Mean height or top height

198. Thickness of bark can be estimated by

(a) Increment borer (b) Swedish bark gauge

(c) Relascope (d) Callipers

199. The volume table gives the volume in round wood down to variable stated thin end diameters is

(a) Commercial volume table (b) Standard volume table

(c) General volume table (d) None of these

200. The instrument used for measuring diameter at unreadable height of standing tree is

(a) Tape (b) Calipers

(c) Dendrometer (d) All of the above

201. The error caused by the instrument, the user and the weather condition is called

(a) Instrumental error (b) Personal error

(c) Error due to observation (d) Error due to leaning of trees

202. Pick the odd one out from the following volume tables classified based on the scope of application

(a) General volume table (b) Regional volume table

(c) Local volume table (d) Sawn out turn table

203. When CAI is equal to MAI

(a) Maximum average volume per unit area per annum is obtained

(b) Average volume per unit area is lowest

(c) The CAI is at its peak

(d) Mai is at its minimum

204. Which of the following gives the correct volume of standing trees?

(a) DBH alone (b) DBH, Height and form factor

(c) DBH and Height (d) Height and form factor

205. Which of the following statement is false?

(a) Volume estimate of individual trees

(b) Estimation of volume of a group of trees

(c) Study the laws of growth

(d) Study stem form of trees

206. Land cover classification using three levels of the scale of 1:100000 was developed by which of the following agency:

(a) IIRS (b) ISRO

(c) NRSA (d) RRSC

207. Remote sensing system which provide their own source of energy for illumination are known as

(a) Optical sensors (b) Active sensors

(c) Passive sensors (d) Thermal sensors

208. Land surface feature has specific response to radiation and it is termed as

(a) Refraction (b) Reflection

(c) Spectral signature (d) Radiation

209. Scattering which occurs when the atmosphere particles are about the same size s the wave length of the radiation

(a) Mie scattering (b) Non-selective

(c) Rayleigh scattering (d) None of these

210. IIRS is located at

(a) Hyderabad (b) New Delhi

(c) Dehradhun (d) Bangalore

211. Ability of the sensors to differentiate different wavelength is known as

(a) Spartial resolution (b) Spectral resolution

(c) Radiometric resolution (d) Tempral resolution

212. Scanning system which collects the data over a range of wavelengths is called as

(a) TM (b) MSS

(c) WIFS (d) None of these

213. Radiation which emitted from the Sun is in accordance with

(a) Kirchoff's law (b) Stefan-Boltzman law

(c) Wien's displacement law (d) None of these

214. Which is the first Indian satellite

(a) IRS-IC (b) IRS-IA

(c) IRS-1 (d) IRS-p1

215. Which of the following wavelength band is used for vegetational studies

(a) Visible only (b) Visible & IR

(c) UV & IR (d) IR, UV

216. Relative brightness of objects is an image is referred to

(a) Texture (b) Tone

(c) Pattern (d) None of these

217. Which of the following is the recently launched satellite with 1m spatial resolution

(a) IRS-IE (b) LANDSAT-7

(c) IKNOS (d) SPOT-III

218. Which of the following is very useful in detecting and mapping forest fires

(a) Visible sensors (b) Thermal sensors

(c) Microwave sensors (d) None of these

219. Aerial photograph which has been taken with optical axis of approximately perpendicular to the horizontal plane & has been deviation upto <40 is

(a) Oblique photograph (b) High oblique photograph

(c) Vertical (d) Low oblique photograph

220. RS & GIS has application in

(a) Biodiversity assessment (b) Vegetation classification

(c) Wild life management (d) All of the above

221. Large scale photograph has scale of

(a) 1:40000 to 1:70000 (b) 1:20000 to 1:40000

(c) 1:50000 o 1:20000 (d) 1:50000 to 1:100000

222. SOI Topo sheet is a

(a) Perspective projection (b) Orthogonal projection

(c) 3D- Projection (d) None of the these

223. NNRMS head quarter in India is located at

(a) Hyderabad (b) Bangalore

(c) Dehradun (d) Ahemedabad

224. Agency for aerial photography in India is

(a) NRSA (b) IIRS

(c) RRSC (d) ISRO

225. High oblique photographs have tilt of

(a) >4 (b) >60

(c) >30 (d) <30

226. Enhanced Thematic Mapper (ETM) has

(a) 7-bands (b) 4-bandsw

(c) 3-bands (d) 8-bands

227. The oldest application of RS technology is in

(a) Agriculture (b) Military

(c) Forestry (d) All of theses

Answer Sheet

1.	(b)	41.	(b)	81.	(a)	121.	(a)	161.	(d)	201.	(a)
2.	(a)	42.	(b)	82.	(b)	122.	(a)	162.	(b)	202.	(d)
3.	(c)	43.	(c)	83.	(b)	123.	(b)	163.	(a)	203.	(a)
4.	(a)	44.	(d)	84.	(a)	124.	(b)	164.	(b)	204.	(b)
5.	(c)	45.	(c)	85.	(a)	125.	(a)	165.	(c)	205.	(a)
6.	(b)	46.	(b)	86.	(b)	126.	(c)	166.	(d)	206.	(c)
7.	(c)	47.	(a)	87.	(c)	127.	(b)	167.	(d)	207.	(b)
8.	(a)	48.	(c)	88.	(b)	128.	(d)	168.	(b)	208.	(b)
9.	(a)	49.	(a)	89.	(b)	129.	(b)	169.	(c)	209.	(a)
10.	(c)	50.	(a)	90.	(c)	130.	(d)	170.	(c)	210.	(c)
11.	(c)	51.	(a)	91.	(c)	131.	(b)	171.	(a)	211.	(b)
12.	(d)	52.	(c)	92.	(c)	132.	(c)	172.	(d)	212.	(b)
13.	(c)	53.	(a)	93.	(a)	133.	(b)	173.	(c)	213.	(b)
14.	(c)	54.	(b)	94.	(b)	134.	(c)	174.	(d)	214.	(b)
15.	(d)	55.	(a)	95.	(c)	135.	(c)	175.	(b)	215.	(b)
16.	(c)	56.	(a)	96.	(d)	136.	(b)	176.	(a)	216.	(b)
17.	(a)	57.	(c)	97.	(c)	137.	(d)	177.	(c)	217.	(c)
18.	(d)	58.	(b)	98.	(b)	138.	(d)	178.	(a)	218.	(b)
19.	(c)	59.	(d)	99.	(a)	139.	(d)	179.	(c)	219.	(c)
20.	(a)	60.	(a)	100.	(b)	140.	(a)	180.	(d)	220.	(d)
21.	(a)	61.	(c)	101.	(d)	141.	(b)	181.	(a)	221.	(c)
22.	(d)	62.	(d)	102.	(c)	142.	(a)	182.	(d)	222.	(d)
23.	(d)	63.	(c)	103.	(b)	143.	(a)	183.	(d)	223.	(a)
24.	(a)	64.	(a)	104.	(a)	144.	(c)	184.	(a)	224.	(a)
25.	(b)	65.	(b)	105.	(d)	145.	(a)	185.	(e)	225.	(c)
26.	(d)	66.	(c)	106.	(a)	146.	(a)	186.	(c)	226.	(a)
27.	(b)	67.	(c)	107.	(d)	147.	(b)	187.	(d)	227.	(b)

28.	(a)	**68.**	(d)	**108.**	(c)	**148.**	(d)	**188.**	(c)		
29.	(a)	**69.**	(c)	**109.**	(a)	**149.**	(d)	**189.**	(b)		
30.	(b)	**70.**	(b)	**110.**	(a)	**150.**	(d)	**190.**	(c)		
31.	(b)	**71.**	(a)	**111.**	(a)	**151.**	(a)	**191.**	(c)		
32.	(c)	**72.**	(d)	**112.**	(c)	**152.**	(b)	**192.**	(d)		
33.	(d)	**73.**	(a)	**113.**	(c)	**153.**	(a)	**193.**	(a)		
34.	(a)	**74.**	(d)	**114.**	(a)	**154.**	(b)	**194.**	(d)		
35.	(d)	**75.**	(c)	**115.**	(a)	**155.**	(d)	**195.**	(b)		
36.	(b)	**76.**	(a)	**116.**	(a)	**156.**	(d)	**196.**	(a)		
37.	(b)	**77.**	(c)	**117.**	(d)	**157.**	(c)	**197.**	(a)		
38.	(c)	**78.**	(b)	**118.**	(d)	**158.**	(d)	**198.**	(b)		
39.	(b)	**79.**	(b)	**119.**	(c)	**159.**	(a)	**199.**	(a)		
40.	(d)	**80.**	(c)	**120.**	(a)	**160.**	(c)	**200.**	(c)		

Glossary

Absolute Increment: The actual increase in tree attributes (e.g., diameter, height) over a specified time period.

Afforestation: The establishment of a forest on land that has not been forested for an extended period.

Allometric Equations: Equations describing the relationships between different tree attributes, such as height, diameter, and volume.

Angle Count Sampling: A method of forest inventory where the number of trees within specified angle intervals is counted to estimate stand density.

Angle Gauge: A tool used to measure tree height indirectly by determining angles and distances.

Arboriculture: The cultivation, management, and study of individual trees in urban and suburban areas.

Area-Based Sampling: A forest inventory method based on sampling plots of fixed size, commonly used in continuous forest inventories.

Bark Characteristics: The texture, thickness, and appearance of a tree's bark, often used for tree identification.

Bark Scaling: Estimation of tree volume by subtracting the bark volume from the total stem volume.

Bark Thickness Gauge: Instrument to measure the thickness of tree bark.

Basal Area Factor Table: A table used to determine the factor for converting basal area to volume.

Basal Area Factor: A constant used in volume estimation equations based on basal area.

Basal Area Increment: The increase in basal area of trees in a stand over a specific time period.

Basal Area Per Acre: The total basal area of all trees per unit area (usually per acre).

Basal Area: Cross-sectional area of tree stems at breast height.

Biltmore Stick: A graduated stick used to estimate tree height and log lengths in the field.

Biomass Expansion Factor (BEF): A factor used to convert stem volume to above-ground biomass.

Biomass: The total weight of living organisms (trees and other vegetation) in a forest area.

Bitter lich Relascope: A sampling tool used to estimate basal area and density in the forest.

Bitter lich Sampling: A method of sampling forest inventory, using a point-centered quarter technique.

Bole Length: The height of the stem of a tree free of branches and defects.

Bole Volume: The volume of the main stem of a tree, excluding branches and foliage.

Browsing Damage: Damage caused by animals feeding on tree seedlings or saplings.

Canopy Closure: The extent to which the forest canopy blocks sunlight from reaching the forest floor.

Canopy Cover: The percentage of the forest floor covered by the tree canopy.

Canopy Gap: An opening in the forest canopy resulting from tree mortality or harvesting.

Carbon Footprint: The amount of carbon dioxide and other greenhouse gases emitted due to human activities, including forest management.

Carbon Footprint: The total amount of carbon dioxide and other greenhouse gases emitted directly or indirectly by human activities.

Carbon Sequestration: The process by which forests absorb and store carbon dioxide from the atmosphere, helping to mitigate climate change.

Climate Change Mitigation: Forest management practices aimed at reducing greenhouse gas emissions and increasing carbon sequestration.

Close-to-Nature Forestry: A management approach aiming to imitate natural forest structures and processes while still producing commercial products.

Cluster Sampling: A method of grouping sample plots in clusters and randomly selecting some clusters for inventory.

Coarse Woody Debris: Dead and downed woody material, such as fallen branches and logs, providing habitat and nutrient cycling in the forest.

Cohort: A group of trees established at the same time and age class within a stand.

Community-Based Forest Management: Forest management involving local communities in decision-making and benefit-sharing.

Composite Sampling: A method of combining samples from different locations to create a representative sample for analysis.

Coniferous Forest: A forest dominated by cone-bearing, needle-leaved trees, such as pine, spruce, and fir.

Coniferous Trees: Trees that bear cones and have needle-like or scale-like leaves, such as pines, spruces, and firs.

Continuous Forest Inventory: An ongoing and systematic process of collecting forest data to monitor changes and provide updated forest information.

Contour Method: A technique for measuring tree heights using a hypsometer and a graduated tape.

Coppice Forest: A forest composed of coppiced trees, usually managed for firewood, poles, or other products.

Coppicing: A regeneration method where trees are cut close to the ground to stimulate regrowth from the stump or root system.

Coppicing: A silvicultural practice of cutting trees to ground level to stimulate new shoots and regrowth.

Cord: A unit of wood measurement, equal to 128 cubic feet or about 3.6 cubic meters.

Crown Base Diameter: The diameter of a tree's crown at its lowest point.

Crown Base Height: The height on a tree where the first live branches are found.

Crown Class Dominance: The level of dominance of a tree within its crown class, such as codominant, intermediate, or dominant.

Crown Class: A classification of trees based on their position in the forest canopy, such as dominant, codominant, intermediate, and suppressed.

Crown Class: Classification of trees based on their position in the canopy, such as dominant, codominant, intermediate, and suppressed.

Crown Competition Factor (CCF): A measure of competition between trees within a stand based on their crown sizes.

Crown Competition Factor (CCF): A measure of tree competition based on crown size and overlap.

Crown Competition Index (CCI): A measure of the degree of competition between neighboring trees in a stand.

Crown Density: The extent to which the tree crowns fill the available space in the forest canopy.

Crown Diameter: The horizontal distance from one side of a tree's crown to the other, typically measured in meters or feet.

Crown Projection Area: The area on the ground covered by a tree's crown.

Crown Ratio Factor: A constant used to estimate crown ratio based on tree height.

Crown Ratio: The proportion of a tree's height occupied by its crown.

Crown Ratio: The ratio of crown length to total tree height, expressed as a percentage.

Crown Thinning: The removal of branches within a tree's crown to improve its health and quality.

Crown Volume: The total volume of a tree's crown, including foliage and branches.

Cubic Scaling: Estimation of tree volume using the cube of the diameter or other dimensions.

Cull Tree: A dead or severely damaged tree that has little or no commercial value.

Cull Tree: A tree that is unmerchantable due to defects or poor quality.

Cumulative Frequency Distribution: A graphical representation of the number or proportion of trees falling within different diameter classes.

Current Annual Increment (CAI): The annual growth increment in volume or other attributes for the current year.

Dead Wood: Standing or fallen trees and branches that are no longer living.

Deciduous Forest: A forest dominated by broad-leaved trees that shed their leaves seasonally, such as oak, maple, and beech.

Deciduous Trees: Trees that shed their leaves annually, such as oaks, maples, and beeches.

Deer Impact: The influence of deer browsing on forest regeneration and vegetation.

Defect: Any physical abnormality or damage in a tree that reduces its commercial value.

Defects: Abnormalities or damages in trees that reduce their commercial value, such as hollows, rot, and forked stems.

Dendrometer: Instrument to measure tree diameter growth over time.

Diameter at Breast Height (DBH): Tree diameter measured at 1.3 meters above ground level.

Diameter at Ground Level (DGL): The diameter of a tree's stem at ground level.

Diameter Distribution Frequency: The frequency of different diameter classes within a stand.

Diameter Distribution Frequency: The proportion of trees in different diameter classes within a stand.

Diameter Distribution Frequency: The proportion of trees within different diameter classes in a forest stand.

Diameter Distribution Frequency: The proportion of trees within different diameter classes in a forest stand.

Diameter Distribution Histogram: A graphical representation of tree diameters within a stand, often displayed as a histogram.

Diameter Distribution: The distribution of tree diameters within a stand.

Diameter Distribution: The pattern of tree diameters within a stand.

Diameter Increment: The increase in tree diameter over a specific time period, often expressed as an annual or periodic increment.

Diameter Limit Cutting: A harvest method where trees above a certain diameter are selectively removed.

Diameter Limit Cutting: A harvesting method where only trees above a certain diameter are cut.

Early Successional Forest: A young forest in the early stages of development after a disturbance.

Ecological Site Classification: The classification of forest sites based on ecological attributes, such as climate, soil, and vegetation.

Ecosystem Services: The benefits and functions that forests provide to human well-being and the environment.

Edge Effect: Changes in environmental conditions and vegetation structure near the boundary of different forest types or land uses.

False Ring: An abnormal growth ring formed due to environmental stress, not representing a true annual growth.

Felling: The process of cutting down trees during harvesting operations.

Felling: The process of cutting down trees during harvesting operations.

Field Data Recorder: A handheld device used to collect and record data during forest mensuration activities.

Field Map: A detailed representation of sample plot locations and forest attributes collected during forest inventory.

Financial Rotation: The optimal rotation age considering economic factors such as timber value, harvesting costs, and interest rates.

Fixed Area Plot: A permanent plot established for repeated measurements over time to monitor stand development.

Fixed Radius Sampling: A method of forest inventory where all trees within a fixed distance from the sample point are measured.

Forest Biomass Carbon: The amount of carbon stored in living trees, roots, and forest floor.

Forest Canopy: The upper layer of a forest formed by the crowns of trees.

Forest Carbon Sequestration: The process of capturing and storing carbon dioxide from the atmosphere in trees and forest ecosystems.

Forest Certification: A voluntary process to verify and ensure sustainable forest management practices.

Forest Certification: The process of verifying and ensuring that forest management practices meet specific environmental and social standards.

Forest Corridor: A linear strip of forest connecting larger forested areas, facilitating wildlife movement and genetic exchange.

Forest Disturbance: Any event that significantly alters the structure and composition of a forest, such as fire, windstorm, or pest outbreak.

Forest Edge Effect: The ecological changes and impacts that occur at the boundary between forest and non-forest areas.

Forest Fragmentation: The division of continuous forest into smaller, isolated patches.

Forest Genetic Resources: The genetic diversity within tree populations, critical for adaptation and resilience.

Forest Grazing: Controlled grazing of livestock within a forest, often used for vegetation management and ecosystem services.

Forest Growth and Yield Models: Mathematical models used to predict tree and stand growth under different scenarios.

Forest Growth Model: Mathematical models used to simulate tree and stand growth over time, considering various factors and management scenarios.

Forest Growth Projection: The estimation of future forest growth and development based on historical data and management scenarios.

Forest Health Assessment: The evaluation of a forest's overall health and vitality, including disease and pest prevalence.

Forest Health: The overall condition and vitality of a forest ecosystem, including tree health and resilience to stressors.

Forest Hydrology: The study of water movement and storage in forested landscapes.

Forest Inventory Plots: Sample areas used to collect data on tree and stand attributes.

Forest Inventory Software: Computer programs designed to aid in data collection, analysis, and reporting in forest inventory.

Forest Inventory: The systematic process of collecting data to assess the quantity and quality of trees and forest resources.

Forest Land Capability Classification: The classification of forest lands based on their potential for various land uses, such as timber production, recreation, or conservation.

Forest Land Use Change: The conversion of forested land to other uses, such as agriculture or urban development.

Forest Mensuration: The scientific measurement and assessment of tree and forest attributes.

Forest Management Plan: A comprehensive document outlining the goals, objectives, and actions for sustainable forest management.

Forest Multifunctionality: The capacity of forests to provide multiple ecosystem services and benefits simultaneously.

Forest Soil Compaction: The compression of soil due to heavy machinery and activities during forest operations.

Forest Stand Structure: The arrangement and distribution of trees in a forest stand, often described using density, size, and species composition.

Forest Stewardship Council (FSC): An international organization that promotes responsible forest management through certification and standards.

Forest Succession: The process of natural development and change in forest structure and species composition over time.

Forest Zoning: Dividing a forest area into different zones or management units with specific management objectives.

Forestry Best Management Practices (BMPs): Guidelines and practices to minimize environmental impacts during forest management activities.

Form Class: A classification of tree shape and form based on stem taper and proportion.

Genetic Diversity: The variety of genetic traits within a population or species, critical for adaptive capacity and resilience.

Genetic Gain: The improvement in desirable traits of tree species achieved through selective breeding and genetic improvement programs.

Geographic Information System (GIS): A computer-based system used for the collection, storage, analysis, and visualization of spatial data related to forests and other landscapes.

Girth: The circumference of a tree stem at a specific height above the ground.

GIS-based Forest Inventory: The use of geographic information systems to integrate and analyze forest inventory data.

Growing Stock: The volume of trees in a stand that is available for harvest.

Growth and Yield: The study and estimation of forest growth and the resulting timber yield over time.

Harvest Scheduling: Planning the timing and intensity of harvesting operations over the long term.

Harvesting Systems: Various methods and techniques used to cut and remove trees from the forest.

Heartwood: The central, older, and darker portion of a tree trunk, often more durable than the surrounding sapwood.

Height Increment: The increase in tree height over a specific time period, typically expressed as an annual or periodic increment.

Height: Vertical distance from the tree base to the top.

Height-Diameter Relationship: The correlation between tree height and diameter, often used for height estimation.

Height-Diameter Relationship: The relationship between tree height and diameter, used for estimating tree height from diameter measurements.

Horizontal Point Sampling: A sampling method that estimates forest attributes based on the number of times a specific point intersects with tree stems within a defined plot.

Huber Value: A constant used in volume estimation for various species based on stem form and taper.

Hurdle Models: Statistical models used to estimate volume when a significant number of trees have zero or below minimum measurable values for some attributes.

Hypsometer: Device for measuring tree height from the ground using trigonometry.

Illegal Logging: The cutting, processing, and trade of timber in violation of national laws or international agreements.

Increment Borer: A specialized tool used to extract tree core samples for measuring annual growth rings.

Increment Core: A small section of a tree core removed with an increment borer for measuring growth rings.

Increment Plot: A long-term forest inventory plot used to measure tree growth and changes in stand attributes over time.

Increment Ratio: The ratio of a tree's height increment to its diameter increment.

Incremental Core Analysis: Examination of tree cores to assess past growth and environmental conditions.

Incremental Growth Ring: The annual ring formed in a tree's trunk, representing one year of growth.

Incremental Growth: The annual addition to a tree's size or volume.

Integrated Resource Management: The coordinated management of multiple resources, such as timber, water, wildlife, and recreation, in a forest area.

Intensive Forest Management: A management approach focused on maximizing productivity and yield through interventions like fertilization and genetic improvement.

Leaf Area Index (LAI): The ratio of total leaf area to the ground area, used to assess forest productivity.

Leaf Biomass: The total weight of leaves in a tree or forest.

Line Intersect Sampling: A method of sampling where the number of intercepted tree stems along randomly placed lines is used to estimate density and distribution.

Log Diameter Class: The classification of logs based on their diameter and quality for different product categories.

Log Rule: A formula used to estimate the volume of logs from tree dimensions.

Logarithmic Rule: A method of volume estimation based on log rules that consider taper and form factor.

Logging Residue: The woody material left on the forest floor after harvesting, including branches, tops, and unused portions of trees.

Long-term Forest Monitoring: Continuous monitoring of forest attributes over extended periods to track changes and trends.

Management Intensity: The level of intervention and control applied to a forest stand for specific management objectives.

Maximal Density Line (MDL): A line representing the maximum density of trees for a given age in a stand.

Mean Annual Increment (MAI): The average annual growth increment in volume or other attributes over a specified period.

Mean Annual Increment in Biomass (MAIB): The average annual growth increment in tree biomass over a specified period.

Mean Square Error (MSE): A measure of the accuracy of an estimation or prediction, calculated as the average of the squared differences between predicted and actual values.

Mean Stand Diameter: The average diameter of trees within a stand.

Mean Tree: An average tree calculated based on sample plot data to represent the stand.

Mensurational Law: A general rule or relationship governing the growth and development of trees.

Merchantability: The quality of a tree or timber product to meet commercial standards and market requirements.

Merchantable Height: The portion of a tree's height suitable for commercial utilization, excluding branches and defects.

Merchantable Timber: Trees or wood products that meet specified quality and size standards for commercial use.

Merchantable Volume: The volume of a tree or stand that meets the quality and size standards for commercial use.

Minimum Diameter Limit: The minimum diameter at breast height required for a tree to be harvested.

Model Assisted Estimation: A method combining statistical models with direct measurements to improve the accuracy of forest inventory estimates.

Mortality Increment: The amount of volume or biomass removed due to tree death.

Mortality Rate: The percentage of trees that die within a stand or population over a specified time period.

Multi-Stem Trees: Trees with more than one main stem originating from the same base.

Multistory Stand: A stand composed of trees of different ages and size classes, forming distinct canopy layers.

Natural Regeneration: The establishment of new trees through natural seed dispersal and germination.

Natural Regeneration: The process of tree establishment and growth without direct human intervention.

Natural Stand: A forest stand that has developed without significant human intervention or management.

Net Annual Increment (NAI): The difference between annual increment and mortality increment.

Non-Timber Forest Products (NTFPs): Goods and resources obtained from the forest other than timber, such as fruits, nuts, and medicinal plants.

Old-Growth Forest: A mature forest with complex structures, diverse species, and little human disturbance.

Overstorey: The upper layer of trees that form the forest canopy.

Overstory: The dominant and tallest trees in a forest stand that form the upper canopy layer.

Pacing: A method of measuring distances in the forest by counting the number of steps taken.

Patch Cutting: A harvesting technique where small patches of trees are removed, mimicking natural disturbance patterns.

Perennial Forest: A forest with tree species that live for many years and reproduce through seeds.

Plotless Sampling: A forest inventory method that does not involve the establishment of predefined sample plots.

Plotless Sampling: A method of forest inventory without predefined plots, using point sampling or line intersect methods.

Point Sampling: A sampling method where the presence or absence of a tree is recorded at randomly selected points within a stand.

Post-Harvest Regeneration: The process of natural or assisted tree regeneration after harvesting.

Precision: The degree of consistency or repeatability of measurements and estimates in forest mensuration.

Pre-commercial Thinning: Thinning of young trees to reduce competition and promote growth before they reach commercial size.

Primary Forest: A forest that has not experienced significant human disturbance or alteration.

Provenance Trials: Research plots established with trees from different geographical regions to assess their performance in a particular site.

Provenance: The geographical origin of seeds or seedlings used in reforestation and afforestation.

Provenance: The geographical origin of tree seed or seedlings used in reforestation.

Pulpwood: Wood harvested for making pulp and paper products.

QMD (Quadratic Mean Diameter): The diameter of a tree with a cross-sectional area equal to the average of all trees in a stand.

Quadratic Mean Diameter: An average diameter used as a representation of tree size in a stand.

Random Sampling: A method of selecting sample plots randomly to avoid bias in forest inventory.

Randomized Complete Block Design (RCBD): An experimental design used in forest research to control variability and test different treatments or management practices.

Reclamation: The process of restoring degraded or disturbed forest land to its natural or productive state.

Recruitment: The process of new trees growing and establishing within a stand, either naturally or through reforestation efforts.

Regression Analysis: Statistical method used to develop equations predicting one variable (e.g., volume) from other variables (e.g., DBH and height).

Relascope: An optical instrument used to estimate the basal area of individual trees.

Relative Density Index: A measure of stand density related to tree size and distance between trees.

Relative Density: The density of trees in a stand relative to the maximum density considered appropriate for the site.

Relative Dominance: The contribution of an individual tree's basal area to the total stand basal area.

Relative Vigour: A measure of a tree's health and growth compared to other trees in the stand.

Release: The process of removing competing vegetation around a desirable tree to promote its growth and development.

Reliability: The degree of accuracy and consistency in measurements and estimates.

Remeasurement: Repeated data collection in a forest inventory to monitor changes over time.

Remote Sensing: The use of aerial photographs or satellite imagery to assess forest characteristics from a distance.

Reproductive Material: Seeds or seedlings used for reforestation and afforestation purposes.

Reproductive Material: Seeds or seedlings used for reforestation and afforestation purposes.

Residual Stand: The trees left after harvesting or thinning operations.

Residual Stand: The trees left standing after a partial harvest operation.

Residual Trees: The trees remaining in a stand after harvesting operations.

Ring Width Analysis: Measurement and analysis of tree growth rings to study past environmental conditions and growth patterns.

Root Collar Diameter: The diameter of a tree's trunk at the base where it emerges from the root system.

Root-to-Shoot Ratio: The proportion of biomass allocated to roots compared to above-ground parts in a tree.

Rotation Length: The number of years between the establishment of a stand and its final harvest.

Rotation Time: The number of years between the establishment of a stand and its final harvest.

Rotational Age: The age at which a forest stand is harvested and regenerated.

Rotational Age: The age at which a stand or tree will be harvested or clear-cut.

Roundwood: Timber in its natural form, as opposed to processed wood products.

Sample Plot: A predefined area where tree and stand measurements are taken to represent the larger forest area.

Sample Tree Selection: The process of choosing specific trees for detailed measurements within a sample plot.

Sample Tree: A representative tree within a sample plot used for detailed measurements.

Sapwood: The outer, living part of a tree trunk responsible for water transport.

Saw Log: A log of suitable size and quality for sawmilling into lumber.

Sawlog Rotation: The number of years a tree or stand takes to reach a desired size suitable for sawing into lumber.

Scaling: The process of measuring and assessing the volume and quality of logs or timber at harvesting sites.

Sectional Survey: A method of forest inventory where a forest is divided into sections for sampling.

Seed Tree: A tree left standing after harvest to produce seeds for natural regeneration.

Selection System: A regeneration method where individual trees or small groups are harvested to create gaps for natural regeneration.

Selective Logging: A harvesting method where individual trees or specific species are harvested, leaving the rest of the stand intact.

Shelterbelt: A row of trees or shrubs planted to provide protection from wind and reduce soil erosion.

Shelterwood System: A regeneration method where mature trees are gradually removed in multiple cuts to create conditions for natural regeneration.

Shelterwood System: A regeneration method where mature trees are gradually removed to create suitable conditions for new seedlings.

Shifted Plot Method: A sampling technique where fixed plots are shifted periodically to monitor changes in tree attributes.

Silvics: The study of the life history and characteristics of forest trees and their interactions with the environment.

Silvicultural Prescription: A detailed plan outlining the management practices to be implemented in a forest area.

Silvicultural System: A planned approach to managing forest stands to achieve specific objectives.

Silviculture: The art and science of controlling forest establishment, composition, structure, and growth to achieve specific management objectives.

Silvopasture: An agroforestry system that combines tree and livestock production on the same land.

Silvopasture: The integration of trees, livestock, and forage crops in a managed forest system.

Site Class: A classification of site productivity based on growth potential for a particular tree species.

Site Factors: Environmental factors influencing tree growth on a site, including soil, climate, and topography.

Site Index Curve Fitting: The process of developing equations or curves to estimate site index based on dominant tree height and age.

Site Index Curve: A graph showing the relationship between dominant tree height and age for a given site.

Site Index Equation: A mathematical formula to estimate site index based on tree height and age.

Site Index Site Quality Classification: The classification of site quality into specific classes based on site index values.

Site Index: A measure of site productivity based on dominant tree height at a specified age.

Site Preparation: Activities undertaken to prepare a site for reforestation, including clearing, soil preparation, and weed control.

Site Preparation: Activities undertaken to prepare a site for reforestation, including clearing, soil preparation, and weed control.

Site Preparation: Activities undertaken to prepare a site for tree planting or natural regeneration.

Site Quality: The capacity of a site to support tree growth and development.

Site Series: A classification of sites based on their potential to support specific tree species or forest types.

Site Suitability Index: A measure of how well a site supports tree growth for a specific species.

Site Suitability Index: A numerical value indicating how well a site can support tree growth.

Site Suitability: The capacity of a site to support the growth and survival of a particular tree species.

Skidding: The process of moving harvested trees from the felling site to a centralized location for processing.

Soil Productivity: The ability of a soil to support tree growth and productivity.

Soil Sampling: Collecting soil samples to assess site fertility and suitability for tree growth.

Spacing: The distance between individual trees in a plantation or managed stand.

Stand Age: The average age of trees within a stand.

Stand Density Control: The regulation of tree density through thinning or other management actions.

Stand Density Index (SDI): A measure of stand density relative to maximum stand density for the site.

Stand Density Index (SDI): A measure of stand density used in forest inventory calculations.

Stand Density Management Diagram (SDMD): A graphical representation of stand density and management options based on self-thinning relationships.

Stand Density: The number of trees per unit area in a stand.

Stand Development Stage: The phase of stand growth and structure, such as establishment, young, mature, and old-growth stages.

Stand Dynamics: The changes in forest structure and composition over time due to natural processes and management interventions.

Stand Dynamics: The changes in forest structure and composition over time due to natural processes and management interventions.

Stand Establishment: The process of creating a new forest stand through planting or natural regeneration.

Stand Height: The average height of trees in a stand.

Stand Heterogeneity: Variation in tree size, species composition, and stand structure within a forest stand.

Stand Projection: Forecasting the future growth and development of a forest stand under different management scenarios.

Stand Table Projection: Predicting future stand attributes using growth and yield models.

Stand Table Projection: Predicting future stand attributes using growth and yield models.

Stand Table Projection: Predicting future stand attributes using growth and yield models.

Stand Table: A tabular representation of forest stand characteristics like tree density, diameter distribution, and volume by size classes.

Stand: A group of trees growing together in a defined area.

Standing Timber: Live trees that have not yet been harvested.

Stem Analysis Chart: A graphical representation of tree growth rings used to determine age and growth characteristics.

Stem Analysis: A detailed examination of a tree's growth rings and characteristics to understand its development history.

Stem Map: A graphical representation of tree stems and their attributes within a stand.

Stem Mapping: The process of collecting detailed measurements of individual tree stems for volume estimation.

Stem Profile: The shape and characteristics of a tree's stem, including taper and form.

Stem Taper Equation: An equation that describes the change in stem diameter along the length of the tree.

Stem Volume: The volume of a tree's stem, excluding branches and foliage.

Stocking: Density of trees in a stand, expressed as a percentage of full occupancy.

Stratified Sampling: A method of dividing a forest area into subgroups (strata) and then selecting sample plots from each stratum.

Stump Height: The height of the tree stump left after harvesting.

Stumpage: The value of standing trees or timber in the forest.

Sub-compartment: A smaller unit of a forest compartment, often used for more detailed management and planning.

Sub-sampling: A method of selecting a portion of the sample plot for more detailed measurements.

Surveyor's Tape: A long, flexible measuring tape used for forest measurements.

Sustainable Yield: The amount of timber that can be harvested from a forest without depleting its productive capacity.

Systematic Sampling: A method of selecting sample plots at regular intervals to ensure even coverage across the forest area.

Taper Model: An equation or function used to describe the reduction in stem diameter with increasing tree height.

Thinning: The removal of some trees in a stand to promote the growth and development of the remaining trees.

Timber Cruising Stick: A tool used in timber cruising to measure tree diameter and height.

Timber Volume: The total volume of merchantable timber in a stand or forest.

Tree Age: The number of years since a tree started growing.

Tree Biomass Allocation: The distribution of biomass within a tree, including roots, stem, branches, and foliage.

Tree Crown Classification: Categorization of tree crowns based on size, shape, and vigor.

Tree Dendrochronology: The study of tree rings and their use in dating and reconstructing past environmental conditions.

Tree Density: The number of trees per unit area in a stand.

Tree Dispersal: The process by which seeds or reproductive material is spread from one location to establish new trees.

Tree Form Factor: A measure of how closely a tree's shape approximates a perfect cylinder.

Tree Marker: An individual responsible for marking trees for harvest or retention based on forest management objectives.

Tree Mortality Factors: Natural or human-induced factors that cause trees to die, such as pests, diseases, and environmental stressors.

Tree Mortality Factors: Natural or human-induced factors that cause trees to die, such as pests, diseases, and environmental stressors.

Tree Quality: The suitability of a tree for commercial use based on its size, form, and defects.

Tree Taper: The rate at which the diameter of a tree decreases with height.

Understorey: The lower layer of vegetation, including smaller trees, shrubs, and herbs, below the main canopy.

Understory: The layer of vegetation beneath the main canopy, consisting of smaller trees, shrubs, and herbaceous plants.

Volume Distribution: The distribution of tree volumes within a stand, often shown as volume classes or bins.

Volume Equations: Mathematical formulas to calculate tree or stand volume based on various measurements.

Volume Ratio: The proportion of the total volume of individual trees in a stand compared to the entire stand volume.

Volume Table: A table showing the relationship between tree dimensions (DBH and height) and their corresponding volumes.

Volume: Amount of space occupied by a tree or stand, often measured in cubic meters or cubic feet.

Volumetric Growth: The increase in forest volume or biomass over a specific time period.

Wedge Prism: A tool to determine tree heights using triangulation principles.

Wildlife Habitat Management: The practice of managing forests to provide suitable habitats for wildlife species.

Wood Quality: The physical and mechanical properties of wood, affecting its use and value.

Yield Table: A table showing the expected yield (volume) of a stand at different ages or stages of development.

9 789359 195001

Printed at Repro India Ltd.